2014
开放式专题设计
OPEN DESIGN STUDIO

清华大学建筑 规划 景观设计教学丛书
《开放式专题设计》编委会 编著

中国建筑工业出版社

图书在版编目（CIP）数据

2014 开放式专题设计 /《开放式专题设计》编委会编著 . -- 北京：中国建筑工业出版社，2016.10
（清华大学建筑规划景观设计教学丛书）
ISBN 978—7—112—17415—7

I.① 2... II.①开 ... III.①建筑设计－研究 IV.① TU2

中国版本图书馆 CIP 数据核字（2016）第 240276 号

责任编辑：陈　桦　王　惠
责任校对：焦　乐　张　颖
书籍设计：Tofu Design（info@tofu-design.com）

清华大学建筑 规划 景观设计教学丛书
2014 开放式专题设计
《开放式专题设计》编委会 编著

*

中国建筑工业出版社出版、发行（北京西郊百万庄）
各地新华书店、建筑书店经销
北京缤索印刷有限公司印刷

*

开本：787×960 毫米　1/16　印张：21　字数：509 千字
2016 年 10 月第一版　2016 年 10 月第一次印刷
定价：**118.00** 元
ISBN 978—7—112—17415—7
　　（29467）

版权所有 翻印必究
如有印装质量问题，可寄本社退换
（邮政编码 100037）

编委会名单

编委会主任
庄惟敏

编委会成员
（以姓名笔画为序）

马岩松　　王昀　　王辉　　齐欣　　庄惟敏

朱锫　　华黎　　李兴钢　　李虎　　张轲

邵韦平　　单军　　胡越　　徐卫国　　徐全胜

梁井宇　　崔彤　　程晓青　　董功

目录 CONTENTS

08	**序言 Preface**	庄惟敏
	开放式建筑设计教学的新尝试	
	A brand-new attempt to the open architectural design teaching	
14	**重建 CCTV 计划**	马岩松
20	勇敢的新央视 /Brave New CCTV	朱钟晖
28	重建 CCTV /Reconstruction CCTV	张博轩
36	珀尔修斯之盾 /Perseus shield	蒋哲
44	猴子的森林 /Monkey Forest	盖郑
45	城市触媒 /Catalyzer of City	金世雄
46	在云端 /Up in the air	邬慧丽
47	天穹花园 /Dome above Garden	谢骞
48	矩阵 /Matrix	徐晨宇
49	在树下 /Under the Tree	张裕翔
50	地下央视 /CCTV Underground	赵健程
51	城市沙堡 /City Sand Castle	周辰
52	魔方 /Cube	朱吴孟健
53	天空之树 /The Sky Tree	邹尚武
54	**理想自宅**	华黎
60	临崖之宅 /House On The Cliff	张文昭
68	双重人格宅 /House for two personality	程正雨
76	三口之家 /House for family of three	谢湘雅
84	巴学园 /Bus Study Garden	陈凌亚
85	迟子建自宅 /Chi Zijian house	范孟辰
86	修行宅 /House for Meditation	高祺
87	墙宅 /House in the walls	郭金
88	音乐之家 /A house for H.O.T	刘涵
89	流浪者的家 /House Of The Nomad	刘奕君
90	天街庐 /House Of Encounter	卢清新
91	山上居 /House On The Hill	王梦熠
92	水上村落 /Micro Village Above Water	朱玉风
94	**生活的舞台**	李兴钢
100	问道——古剑奇谭游戏 / 人生体验中心 /Asktao	刘楚婷
108	微社区 /Micro Community	逄卓

116	梦工厂 /Dream Factory	蔡宙燊
117	微城市 /Micro City	窦森
118	变形虫 /Amoeba	金容辉
119	猫咪咖啡馆 /Cat Cafe	田莱
120	艺圃 /Garden Art	王冉
121	心灵院 /Mind House	余乐

122　城乡聚落　　　　　　　　　　　　　　　　　　张轲

128	假想未来 /3D City	殷玥
136	1KM-HIGH 垂直聚落 /Vertical Village	沈毓颖　张孝苇　李培铭
144	折叠的风景 /Folded Landscape	马宁　蔡忱
152	在生活中修行 /Practice In Life	张乃冰
153	垂直村落 /1KM—HIGH Vertical Habitat	许东磊　徐滢
154	飘 /Lightweight Network	刘东元　杨良崧
155	极细聚落 /A Thin Habitat	陈晓东

156　建筑师之家　　　　　　　　　　　　　　　　　　邵韦平

162	PLASMA 建筑师工作营 /Architect Work Camp	马逸东
170	对望·芦苇荡 /Look, the reeds	孙冉
178	三合院 /Courtyard	刘畅
186	筑·园 /Architect Park	刘凤逸
187	建筑师周末俱乐部 /Architects weekend Club	王之玮
188	建筑师之家 /Architects'home	康凯
189	穿行 /Through	杨烁
190	建筑师之家 /Architects'home	段俊毅
191	建筑师之家 /Architects'home	郑松
192	建筑师之家 /Architects'home	黄河
193	隐·居 /Hidden home	徐珂

194　通过限制创造 " 无限 " 的可能　　　　　　　　　　胡越

200	艺术家工作室 /Artist studio	甘旭东
208	视界 /Horizon	孙鹏程
216	根雕艺术家工作室 /Carving artist studio	李旻华
224	MW 版画家工作室 /Version Of The Artist Studio	陈羚琪
232	浮岛 /Floating Island	张启亮
233	地平线 /Horizon	崔永

234	装置艺术家工作室 /Installation Artist Studio	吕代越
235	夫妻工作室 /Husband and wife studio	马步青
236	空中温室 /Air greenhouse	齐大勇
237	雕塑之家 /Sculpture House	谯锦鹏
238	艺术孵化器 /Art incubator	谭婧玮
239	光影艺术家工作室 /Shadow artist studio	王靖淞
240	有限空间的无限组合 /Infinite combination of finite space	吴珩
241	艺术家工作室 /Artist studio	杨茜

242 日常经验·个体记忆 + 居住畅想　　　　梁井宇

248	桌 /House With A Table	刘潇潇
256	上下之间 /Family House	赵慧娟
264	依树居 /Living By Trees	唐紫晔
272	共享书宅 /Reading—Home	黎雪伦
280	卧居 /Sleeping—home	戴锐
281	猫意栖居 /Living like a cat	刘畅
282	回归自然·消融边界 /Boundary Ablation	李嘉雨
283	小型悬挂住宅 /Suspended House	刘少文
284	寻回童年的院子 /Courtyard for Childhood	薛昊天
285	里外 /Interior& Exterior	闫博
286	倚墙而居 /Living Against The Wall	袁雪峰
287	美食·家 /House From Food	周翘楚

288 "光的空间" 建筑设计　　　　董功

294	模糊 /vague	王杏妮
302	柱光亭 /Meditation Pavilion Wudaokou	张成章
310	地球的子宫 /Earth Womb	秦正煜
318	光的呼吸 /Breath of Light	葛肇奇
326	光影交响 /Symphony of light	辛大卫
327	天·地·光·城市 /Sky, earth, light, city	张若诗
328	水·星·光 /Water, stars, light	赵萨日娜
329	光的空间 /Space of light	杨宝诚
330	树之眼 /Eye of the tree	周祎馨
331	光的冥想空间 /Light meditation space	左碧莹
332	光的滤镜 /Optical filter	唐宁
333	时·光·序列 /Time, light and sequence	李新新

334　　　致谢

序言 PREFACE

开放式建筑设计教学的新尝试
A BRAND-NEW ATTEMPT TO THE OPEN ARCHITECTURAL DESIGN TEACHING

庄惟敏 Zhuang Weimin

开放式建筑设计教学缘起于以下几点：第一，是与我们建筑学教育的特点有关，建筑学的学科特征从本质上来讲，是伴随着人类社会的进程而发展的，它是一个人居环境的营造过程，把这个过程中的教授与传承演变成为一个教化人的教育过程，就成为今天的建筑专业教育。师徒相传是建筑教育最本质也是最朴素的方法。无论是中国的鲁班，还是美国西塔里埃森的赖特，他们的专业传授中不仅有知识的传承，更重要的是作为一个建筑师的身体力行的影响。由于我们学制时间的限制，怎么样能使学生在有限的时间内，尽快变成一个对建筑学专业有理解、对未来自身定位有明确的方向、尽快进入职业状态的职业建筑师，这一点是很重要的。

第二，是学校教育层面的思考。在清华大学第24次教学讨论会的开幕式上，校领导明确指出，改革开放30年来清华的教育是值得反思的：一是优秀的学生能进来不能保证他们优秀地出去，或者说有没有做到更优秀；二是学生在清华有没有培养出自己对专业的志趣。志趣没有，就看不到未来作为一名职业建筑师的状态。环顾我们周边的新一代建筑师，他们在建筑实践的第一线，无论是实践作品还是思想理念都非常精彩，已经具有了很大的社会影响和业界赞誉，他们的执业状态是可以影响年轻人的。这一点对学生们来说更重要。这些中青年建筑师有各自的建筑理想，有积极的生活态度，作为一名活跃的建筑师，他们会有很多的感悟，但同时他们又身体力行，让他们来教授学生们，让学生们感受他们对建筑的热爱和执着，远远胜过书本里的说教，这种直观形象的感受是最感人的。

第三，是从职业角度的思考。我本人从2005年开始到现在一直在UIA职业实践委员会工作，这么长时间以来特别明显地感到，我们现在的职业教育和国际上还有很大的差距，这个差距不是说我们课程设置怎么样，也不是说老师有怎么样的问题，最主要的是我们缺乏一种职业精神。我们的本科和硕士研究生教育是职业教育，我们培养的建筑学的学生是要拿专业学位的，是为了以后成为职业建筑师的。但针对这一目标，即我们的学生是否毕业以后能够达到一个职业建筑师的起码要求，我还是担心的。因为，按照以往的教学模式，作为职业建筑师的基本职业素养和职业精神，往往是在学校里学不到的。所以必须要有人手把手地教，这些人是需要经历过市场的，经历过很严酷竞争的，经历过和社会面对面、和业主面对面、和批评家面对面交流的，在他们身上反映出来的职业素养，其实就是最好的职业建筑师的表率，就是最好的职业精神的表现，他们的职业精神传授就是最好的教科书。

基于以上三点，寻求开放式教学的缘由变得愈发充分。正好有这样一个契机，2013年的北京市优秀建筑设计评审，北规委邀请了除资深专家和院士之外的一批中青年建筑师，他们有一些是民营建筑设计企业的总建筑师，有些是独立建筑师事务所的领衔建筑师。这些建筑师都在创作的第一线，都有敏锐的专业思考，同时还有很好的专业背景和国际视野，都有海外留学的经历。他们的参与对北京市优秀建筑的评优在学术层面起到了绝对的推动作用。在那次评审会后，我就更加坚定了这个想法，要聘请他们作为清华设计教学的导师，用他们的亲身实践和职业精神影响到学生。经院务会研究决定，由学院主管教学的副院长单军教授组织院教学办制定计划，第一批确定了15位中青年建筑师作为三年级设计课的导师。然而情况并不顺利，按照清华大学的一般做法，是没有"设计导师"这一聘任头衔的。于是，我们向学校专门陈述我们的理由和想法，讲述国外"评图导师"的先例。在我们向陈吉宁校长汇报设想之后，得到了校长的肯定，校人事和教务部门也对此给予大力的支持。15位导师的证书是由清华大学颁发的"清华大学建筑学设计导师"聘书。这件事情就这样开始了。

最初开始时，我们对这件事的目的和意义并没有领会得太深，但在后来的教学和若干次的教学讨论的过程中，发现有两方面情况：一是这些设计导师们都很有荣誉感和责任感，很有激情，愿意用他们的经验去教导学生；二是，这些建筑师毕业于中国现有的教学体制，大部分又受过西洋教育，有宽广的视野，他们有自己的判断，这个判断是对当下我国建筑教育的判断，很多人直言不讳地对当下的建筑教育提出了批评，许多是以他们亲身经历的感悟而提出来的。这样我更觉得这件事情是有意义的，它不是一个简单的review，或者final review，而是希望他们参照自己以往的学习经验，在新的过程中教会学生去思考。同时，我也希望这些建筑师的引进可以触动我们自己的老师，我们的老师还有一些是纸上谈兵的东西，或者就理论说理论的东西。能盖出好房子的老师在业界是公认的好老师，至少他有实践经验，如果他的建筑又得了奖，那么他在学生们的心目中也是备受瞩目的，这就是设计导师的力量。

我们没有像其他学校那样设立"大师班"，而是在一个年级全面铺开，全面铺开的优越性在于：首先让学生们全盘面对一个真实建筑师的舞台，这些建筑师是与国际建筑师在同一个舞台上对话的，这一点让学生们很兴奋；其次是这些建筑师可以把他们想的东西讲出来，这种讲述，对自身而言也是一种互动；第三也是比较关键的，我们希望尝试一种教学方式的变化，这种变化的出发点，都是希望我们能有一个相对务实的、同时又很有效率的方法，来启发和教育学生对专业学习的志趣。

设计导师们会把自己作为职业建筑师的想法和体验灌输给这些年轻的学生，使学生们强烈地感受到来自职业建筑师的信念和追求。这一点恰恰是我们当下设计课教学中最缺乏的，也是学生们最需要的。各位导师出不一样的题目，都紧扣了当下建筑设计的热点，以及建筑教育的关键点，而学生们的自由选择导师和题目的过程，也反映了我们学生们的价值取向，能够大胆地按照自己的选择去表达。

这件事另外一个附带的效果是仪式感。仪式感对建筑学的教学是很重要的，国外很多建筑院校的评图是相当有仪式感的，我们需要用仪式感去调动学生的主观能动性和成就感。这场秀一定要做，是在学生们面临市场和社会选择的情况下秀自己，同时也是秀老师。仪式感在公开评图那天相当起作用，因为建筑本身就是带有很强自我表现的科学和艺术的综合体，是一种品质的体现，综合气质的展现。

开放教学已经进行快两年了，设计导师们的激情不减，学生们的热情不减。每一次的教学研讨和学期初的教学准备会我们都会听到导师们对教学的思考和再认识，也一次次地被这些设计导师们的职业精神所感动。每一次的教学评图，无论是中期还是期末，我们都会听到学生们充满激情的反馈，也一次次地感受到学生们对这种教学模式的喜爱。我们为此感到欣慰。

希望我们的同学们可以更积极地参与其中。我们期待着明年。

The idea of open architectural design teaching originated for the following causes: first of all, it is related to the characteristic of architecture education. The discipline develops along with the development of human society since the beginning, which is a building process of living environment. Teaching and passing on knowledge during this particular process forms today's architecture education. The most substantial and simple teaching method is the apprenticeship. Being either Luban from China or Frank Lloyd Wright known for Taliesin West, the passing on of his expertise includes not just the passing on of knowledge but to influence as an architect figure. Due to the limited length of school education, how to provide students with qualified education that would ensure they develop into professional architects with deep understanding of architecture expertise and clear vision of future in such a limited period of time is a vital theme.

The second source of the idea is, the reflection on college education. On the opening ceremony of the 24th teaching forum of Tsinghua University, the university leader has pointed out clearly that the education of Tsinghua University, for three decades since the Founding and Reform, is worthy of reflection: on one hand, the top students entered with no guarantee of their excellence when graduating, or to say not exceeding; on the other hand, whether Tsinghua has nurtured the interest of the students in their major or not was unclear. Without interest, one could not see himself as a professional architect in the future. Retrospectively, architects of the new generation stand in the front-line of architecture practice. Being brilliant both in theory and practice, their practice of work and their minds will influence the society and form their reputation.. Their working status could impact the younger generation. This is of great importance to the students. All of the established architects or young architects have their own architectural dream as well as positive living attitude. As an active architect, they must have much feeling yet practice by himself. Allow them to teach the students and encourage the students to search for their enthusiasm and persistence on architecture prevail over the textbook lessons. The vivid and live experience is the most touching.

Thirdly, thinking from the occupational perspective. Since 2005, I have been working at UIA Professional Practice Commission. For a long time, I have noticed a very obvious fall out between the level of our domestic professional education and the international level. The gap exists in neither the course setting nor the teachers, but rather the lack of professional spirit. Our undergraduate and graduate education is parts of professional education. The students we nurtured are to get professional degree and become architects. I hold a little concern toward whether our students could live up to the requirements for a professional architect. As for the previous teaching mode, the basic professional spirit and virtues could not be taught at school. There must be hand by hand tutoring by those who have survived from the fierce competition in the market and experienced the face-to-face communication with the clients and the critics. The professional virtues in these people are the best examples as professional architects.

Based on the three aforementioned aspects, there are sufficient reasons to pursue open teaching. There appears to be a great opportunity then when the 2013 Beijing Best Architecture Design appraisal invited some newly established architects and young emerging architects besides senior experts and academicians. They are the chief architects from private architecture design enterprises or founding architects from independent firms. All of them are from the front-line of innovative creation with clear visions, excellent expertise background and global perspectives rooted from their overseas education experience. Their participation in the appraisal definitely propels the academia. After the appraisal, I am convinced to hire them as teaching instructors of architecture design for Tsinghua to influence the students with their practical and professional spirit. The faculty commissioned the deputy dean Pr. Shanjun, who is in charge of teaching, to organize and set up the course. A first batch of 15 newly established architects and young emerging architects were to become tutor for junior-year teaching course. However, the process is not so smooth. According to the normal practice of Tsinghua University, there is no such title as "design tutor". Thus, we specially stated our reasons and thoughts to the university and the precedent of "review tutor" adopted abroad. Our report to President Chen Jining received approval and great support from the HR and Teaching Department. The certificates of Tsinghua Architecture Design Tutor for our 15 tutors was issued by Tsinghua University. Thus the whole story began.

At the very beginning, we haven't fully grasped its purpose and significance, but during the post course review and discussion, we noticed the two situations: First, these architects carried great sense of honor and responsibility, who are passionate and willing to teach students with the extent of their capacity; Second, though the architects are educated under the typical teaching system in China, many of them also have the experience of studying abroad with broad horizon and independent judgment. Their critics are made against the current architecture education, and even blunt criticism from their personal experience. I believe this is significant. It is not a simple review or final review, but that they want to refer to their previous study experience and teach the students to think during the learning process. Meanwhile, I also hope that the introduction of the architect could influence us as teachers, who talk in theory with no practice. The best teacher is the one who could build great houses. At least he should have some practice experience. If his building was publicly rewarded, then he became a mentor among students. This is the power of architect tutor.

Unlike the "Master's Course" in other universities, we offer equal opportunity for all students in the same year to participate in the course taught by architecture masters. The advantage of this widespread studio is that: first students are confronted with influential practicing of architects, who is on the same platform as international architects, which the students are thrilled about; secondly, the architects could talked and rephrase their thoughts and ideas which become an revision for themselves; last but not least, we hope to try a different teaching mode. This change is our wish to have a more practical and yet efficient method to raise the interest of students and educate them towards professional study. Architect tutor would pour their professional ideas and experience into the young students bring about the strong belief and dream as a professional architect. This is what we lack the most in our current design teaching courses and what the student need the most. Each tutor has a different topic, firmly focusing on the current hot issue of architecture design and the key points of architecture education. Whilst the free selection of tutor and topic, the student's choice of selecting reflects the value orientation of our students and encourage them to boldly express themselves based on their own choice.

Another effect is the sense of ceremony, which is very important to architecture teaching. In many architecture colleges overseas, the review process is very ceremonial. We need the sense of ceremony to activate the students' subjective initiative and sense of accomplishment. The show must go on. Particularly, the students could display themselves and be confronted by the selection by market and society, and demonstrate to the teachers as well. The sense of ceremony worked really well on the Public Review Day, since architecture is an integration of science and art with strong need for self-expression. It is also a demonstration of the overall ability.

It has been almost two years since we introduced open teaching. The design tutors are passionate as ever and so do the students. For each teaching discussion and the teaching preparation at the beginning of the term, I would listen to the tutors about their thoughts and reform my understanding of teaching. Repeatedly, I was moved by the professional spirit of the design tutors. Every time at the design review, either mid-term or final, we would receive warm feedback from our students. Over and over, we see the students are affected during this teaching mode. We are genuinely gratified with what we have accomplished.

Hope my students could participate more actively.

Looking forward to the coming year.

重建 CC

马岩松

MAD 建筑事务所创始人、合伙人

TV 计划

马岩松
MAD建筑事务所创始人、合伙人

教育背景
1994年 – 1999年
北京建筑大学建筑与城市规划学院建筑学 学士
2000年 – 2002年
美国耶鲁大学建筑学院建筑学 硕士

工作经历
2004年至今
MAD建筑事务所创始人、合伙人

主要论著
Mad Dinner [M]. Actar出版社, 2007
《疯狂兔子》[J]. a+u, 2009/12
10*10/3[M]. Phaidon出版社, 2009
Douglas Murphy. "It's a MAD World" [J]. ICON, 2011/04
Edwin Heathcote. "Conquering the West" [J]. 英国金融时报, 2011/03
Matthew Allen. 封面报道 "An Empathetic Twist" [J]. Domus, 2012/11
马岩松. Bright City, Blue Kingfisher出版社, 2012
Ma Yansong From (Global) Modernity to (Local) Tradition, Actar出版社 & FUNDACIÓN ICO, 2012
马岩松.《山水城市》, 广西师大出版社理想国, 2014
马岩松. Shanshui City, Lars Müller出版社, 2015
马岩松.《鱼缸》, 中国建筑工业出版社, 2015
MAD WORKS, Phaidon出版社, 2016
Isabelle Priest. 封面报道 " Stepping Forward, Looking Back" [J], Riba Journal, 2015/12
Alexandra Seno. 封面报道 " Dangerous Cures" [J]. Architectural Record, 2015/12
Sara Banti. 封面报道 " A New Harmony between Art and Nature" [J]. Abitare, 2015/12
Harry den Hartog. 封面报道 " Taking Nature to the Next Level" [J]. Mark, 2015/12
马传洋. 封面报道《马岩松MAD》[J]. PPaper, 2016/01
封面报道《哈尔滨大剧院》[J]. Domus中文版, 2016/02
封面报道《哈尔滨大剧院》[J]. 世界建筑, 2016/02
方振宁. 封面报道《建筑与自然和音乐迂回》[J]. 时代建筑, 2016/03
杨志疆.《第二自然的山水重构》[J]. 建筑学报, 2016/06

设计获奖
2008 – 全球最具影响力20位青年设计师, ICON
2009 – 全球最具创造力10人, Fast Company
2011 – 国际名誉会员, 英国皇家建筑学会
2011 – 中国最具创造力10公司, Fast Company
2012 – 黄山太平湖公寓: 2012年度十佳概念建筑, Designboom
2012 – "假山": 最佳住宅群建筑, 世界地产奖
2012 – 梦露大厦: 年度建筑, ArchDaily
2012 – 梦露大厦: 美洲地区高层建筑最佳奖, CT BUH (高层建筑与人居环境委员会)
2013 – 梦露大厦: 2012全球最佳摩天楼, EMPORIS
2014 – 2014年度世界青年领袖, 世界经济论坛
2014 – 全球商界最具创造力100人, Fast Company
2014 – 鄂尔多斯博物馆: 最佳"建筑-金属"奖, 世界建筑新闻奖
2014 – 南京证大喜玛拉雅中心: 2014年度十佳住宅, Designboom
2014 – 朝阳公园广场: 2014年度十佳高层建筑, Designboom
2015 – 威尔士大道8600: 设计概念奖, 洛杉矶建筑奖
2015 – 康莱德酒店: Beyond LA奖, 洛杉矶建筑奖
2015 – 全球设计权力榜, Interni杂志
2015 – 哈尔滨大剧院: 2015年度十佳艺术中心, Architectural Record
2015 – 哈尔滨大剧院: 2015年度最瞩目建筑, Wired
2016 – 哈尔滨大剧院: 年度文化建筑, ArchDaily
2016 – 哈尔滨大剧院: 最佳"表演空间"奖, 2016世界建筑新闻奖
2016 – 哈尔滨大剧院: Beyond LA奖, 洛杉矶建筑奖

代表作品
加拿大梦露大厦(图1)、哈尔滨大剧院(图2)、鄂尔多斯博物馆(图3)、日本四叶草之家(图4)、芝加哥卢卡斯叙事艺术博物馆(图5)、中国木雕博物馆(图6)、北海"假山"(图7)、胡同泡泡32号(图8)、黄山太平湖公寓(图9)、北京朝阳公园广场(图10)、南京证大喜玛拉雅中心(图11)

重建CCTV计划

三年级建筑设计（6）设计任务书
指导教师：马岩松

计划前言

关于CCTV新大楼的争议有关于美学的，关于建筑结构和造价的，也有关于这幢大楼的社会性，文化性，和政治性的。一个建筑项目如此受到关注，也说明了这样的建筑在现实的中国是要面对考验的，这也是中国城市化进程要面对的真正挑战。来自中国的纷杂的声音让我们不禁要问，有多少的争议是真正有建设性的？北京的CBD应该如何规划，大体量的城市建筑将面对什么样的未来？如果不是现在的CCTV，我们理想中的未来大城市是什么样子？还是我们根本就没有过这样的理想？

我们把这个问题抛给学生们，这些未来的建筑师们，他们应当怎么去回应这样的问题？如果把这个项目交到这些未来的建筑师手里，又会是怎样的一个结果？

思路梳理

历史性反思

对宏大建筑的近现代史进行梳理。对早期苏维埃纪念碑式建筑和20世纪初美国的权力资本象征的摩天楼做对比性研究。从形式的差异追究意识形态的差异，进行历史层面的批判和反思，以解构建筑意识形态为目的。

城市如何回归人性？

建筑不是居住的机器。自现代主义以来的城市和建筑最显著的问题就是人性的缺失。城市应该以大部分人的整体利益为前提，而不应该成为被政治和资本操控的机器。人作为城市的主体，城市的建造者和使用者，应该处在怎样的一个位置？

城市精神从何而来?
对现代主义代表性的纽约曼哈顿和老北京进行对比分析,发掘东西方城市在气质上和精神上的不同。对中国城市的传统价值进行再发掘和再认识。讨论中国城市的精神理想。

重建CCTV
人们还从来没有去想象过推翻已经存在的现实,充满理想主义的城市计划是当代中国所需要的,"少一些美学,多一点伦理"是本课程所期待的。

理想城市
盲目的城市化进程仍旧在中国继续推进,我们迫切需要新的城市发展思路去扭转当下的局面。而CCTV这个建筑群影响着CBD的规划,对北京城市的生活、文化、生态等方面都有不同程度的影响。针对复杂的大建筑城市系统的认识如何得到批判性的,诗意的表达?针对建筑还是整体环境?

课题内容
前半学期:进行计划第二部分的"思路梳理"和对"理想城市"的讨论。
后半学期:对计划进行实质性操作,并阶段性汇报和交流。

课题表达形式
模型实体,图纸,也可有动态影像,平面手绘。

课题成果展示
最后的成果将和CCTV投标时的其他方案以群展的方式展示,并请当时的竞标评委出席展览和评图。

勇敢的新央视 /
BRAVE NEW CCTV

项目选址：北京央视大楼原址
项目类型：媒体中心，城市地标
建筑面积：50 万 m²
用地面积：20 万 m²

方案设计：朱钟晖
指导教师：马岩松
完成时间：2014

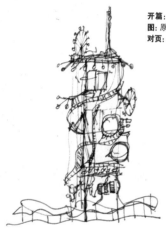

开篇：北部入口人视。**本页上图、左图**：原始概念。**本页下图**：功能地图。
对页：主要方案图。

在思考未来的媒体时，两本书对我产生了很大的影响，一本是奥威尔的《1984》，另一本是赫胥黎的《美丽新世界》。反观现状，当1984年真的到来时，预想的专制独裁早已被打垮，而娱乐至死的社会仿佛真的成了现实。这次设计的出发点正是对这种娱乐至死的社会现象的批判。用游乐场的意向，将建筑转化为游乐场，将CCTV的空间、功能与游乐设施相结合，变成了一个永远转动的、人声鼎沸的甚至疯狂的游乐场。所有人都可以来到其中获得娱乐，而内部的办公人员也在游乐场般疯狂的办公场所中工作。

然而批判应该留给文学家，建筑师在批判之后更应该有所建树，寻找出路。建筑师的使命不是在城市的伤口上撒盐而应该努力治愈伤口。游乐场的想法让封闭的CCTV向城市和市民开放，邀请所有人参与，在单调甚至无聊的CBD中创造出一片轻松愉快的场所，为CBD注入活力。设计从批判的角度出发，得到一个治愈的结果。

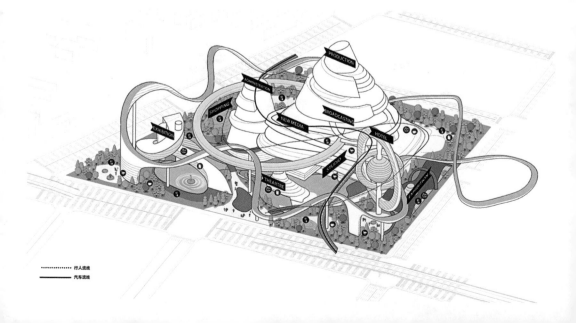

In thinking about the media of the future, two books have had a great impact on me, one is Orwell's 1984, and the other is Huxley's Brave New World. On reflecting the present situation, when 1984 really comes, the expected dictatorship has been defeated, and the society of entertainment to death seems to be real. The starting point of this design is a criticism to this social phenomenon of entertainment to death. With the intention of the playground, the building is transformed into a playground. It becomes a permanently rotating, crowded and even crazy playground with the combination of CCTV space, function and

recreation facilities. All the people can come to have entertainment, and the internal office staffs also work in the crazy office as playground. However, criticism should be left to writers, and architects should only accomplish achievements after confronting criticism and find a way out. The mission of an architect is not to rub salt in the wound of the city, but should strive for healing. The idea of the playground opens the otherwise closed CCTV to the city and the public, inviting all people to participate, create a relaxing and pleasant place for the monotonous or even boring CBD, by injecting vitality to CBD.

上跨页：故事分布。**本页上图**：功能重组。**本页下图**：南部入口人视。**对页上图**：室内透视。

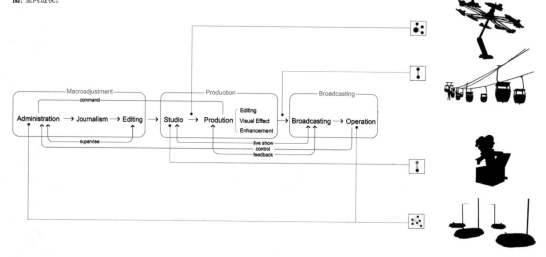

教师点评

这是一个既诙谐又深刻的作品:央视大楼功能上的复杂和特殊性演绎成了不同于千城一面、虚伪的、表象的、新的类型的城市,娱乐性的隐喻也让那些貌似一本正经的宏大叙事般的纪念式建筑变得滑稽。她重新检验了建筑组织和形式在城市中的象征式意义。

Teacher's comments

It is a witty and profound work: The complexity and particularity in function of the CCTV building interprets a new type of city that is different from other hypocritical cities with being superficial and lacking individual characteristics. The entertaining metaphor also makes those priggish, grand-narrative, and monumental buildings funny. She reexamines the symbolic meaning of building organizations and forms in cities.

重建CCTV /
RECONSTRUCTION CCTV

项目选址：北京央视大楼原址
项目类型：媒体中心，城市地标
建筑面积：50 万 m²
用地面积：20 万 m²

方案设计：张博轩
指导教师：马岩松
完成时间：2014

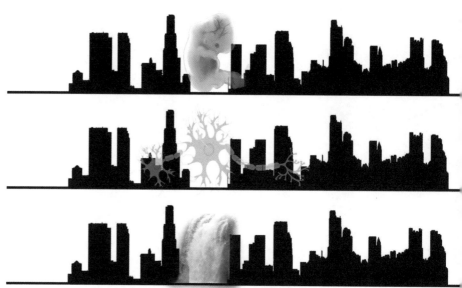

库哈斯在《Delirious New York》一书中将城市消解为个体价值的实现，表明了对于个体的关注，然而却在实践中将"个体"的概念泛化为商业时代具有强大力量的机构、组织，它们可以被称作个体，却不具有每个人作为个体的孤独脆弱的性质。而媒体，作为人与人之间的连接，应当重新将关注点集中到每一个真正的个体——人，以及个体间的联系。作为新一代媒体建筑的央视大楼将像一个或一组神经元那样，不断和外界交换信息，联系每个个体，成为整个城市的神经中枢。

人不能两次踏入同一条河流，这是自然的魅力和伟大之处。商务核心区，原本是属于钢铁巨兽的领地，它们冰冷、无情、永恒。作为一个城市景观，矗立于北京CBD的央视大楼将是一种自然的、随机的、混沌性的表达，如同瀑布般永恒变化，以瞬息万变对话永恒。

开篇：方案主要透视。**对页上图及本页上图**：概念分析。**对页下图及本页下图**：主要模型透视。

库哈斯的CCTV无疑是成功的。库哈斯用他敏锐的洞察力为北京带来了新时代需要的建筑，使之从一个建筑本身升级为一个事件。然而当我们这样认为时，我们是不是已经不知不觉被他强硬的态度和建筑语言所压迫？

库哈斯在《CONTENT》一书中自豪地将新CCTV称为Beijing Manifesto，现代主义有了宣言，曼哈顿有了宣言，北京一定要有宣言吗？宣言就像是一个句号，总是极力地归纳、截取、结束。然而作为一个城市，北京在发展。北京面临数不清的问题。与其他大都市相比，她更像一个早熟的婴儿，疾病缠身却顽强生长。她混沌、脏乱、安详、美丽。她需要的是否是个句号？是否禁得起一个句号的分量？她又岂是一段宣言所能概括？

与现代主义的宣言式的表达相对，我更希望通过一种东方式的、叙事性的方式，描绘出一个脆弱的生命体的形象，它就好像一个胎儿，在城市中孕育、生长，柔软、脆弱而敏感，代表了这个城市的混沌、未知以及对于未来的开放。她看似柔弱，然而所谓"润物细无声"，作为一个地标建筑，她所能传达出的，难道会弱于那些斩钉截铁的宣言？

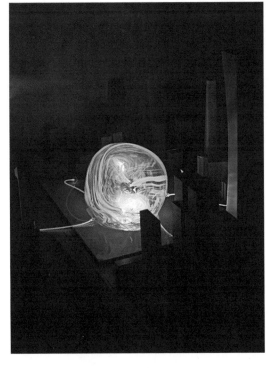

Koolhaas' CCTV New Building is no doubt a success. Koolhaas bought the building needed in the new era to Beijing with his keen insight, upgrade it from a building itself into an event. However, when we think of it, have we been oppressed by his tough attitude and architectural language already? Koolhaas was proud to call CCTV New Building as Beijing Manifesto in the book, CONTENT. Modernism has a declaration, Manhattan has a declaration, and must Beijing have a declaration? Declaration is like a full stop, which is always trying to make summarization, interception and ending. However, as a city, Beijing is developing. Beijing faces numerous problems. Compared with other big cities, she is more like a precocious baby, full of the diseases but still tenacious to grow. She is dirty, chaotic, serene and beautiful. Does she need a full stop? Can it bear the burden of a full stop? How can she be summarized into a section of declaration?

Opposite to the declarative expression of modernism, I hope to depict the image of a fragile life through the Oriental style and narrative method. It is just like a fetus, being conceived and growing in the city. It's fragile and sensitive, which represents the chaos, mysteries of the city and openness to future. Although she seems weak, she is what the Chinese saying "moisten things silently" goes. As a landmark building, should what she can convey be weaker to the emphatic declaration? Koolhaas decomposed the city into the realization of individual value in Delirious New York, indicating the concern for the individual. But in practice, he generalizes the concept of "individual " into the institution and organization that have strong power in the commercial era. They can be called as individual, but they don't have any of the the lonely and fragile nature as an individual. And the media, as a connection between people, should focus on every real individual again - people, as well as link between individuals. CCTV New Building as a new generation of media building, like one neuron or a group of neurons, will constantly exchange information with the outside world and contact each individual, becoming the nerve center of the whole city.

本页上图: 鸟瞰图。**对页下图**: 剖面图。

教师点评

中国的城市CBD将西方的工业文明下的城市价值推向了新的极致——更高、更快、更强,造就了一个个坚硬、富有力量,却缺少人性和情感的城市和建筑,包括已经建成的CCTV大楼。张博轩的作品《孕育》像一个柔软、敏感的胚胎,在强硬的环境中存在,像一种温柔的抵抗。它是存在于互联网信息时代,代表着新文明、新思想的重要作品;无论出发点还是语言都和现代城市截然不同。她好像带着呼吸,时刻提醒我们,城市中人的存在。

Teacher's comments

The CBDs in China push the urban value under the industrial civilization of western countries towards a new extreme — higher, faster and stronger, forging various solid and powerful cities and buildings lacking humanities and emotions, including the CCTV building. Zhang Boxuan's work Breeding is like a soft and sensitive embryo that exists in the tough environment, just like a kind of gentle resistance. Existing in the Internet information age, it is a significant work representing the new civilization and thought, which is quite different from the modern city in terms of the starting point and language. It seems to have the breath and always reminds us the existence of humans in the city.

本页上图：夜景效果图。**本页下图**：室内效果图。

珀尔修斯之盾 /
PERSEUS SHIELD

项目选址：北京
项目类型：媒体建筑
建筑面积：1km²

迪拜

纽约

北京

香港

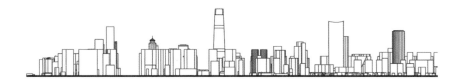

方案设计：蒋哲
指导教师：马岩松
完成时间：2014

开篇: 俯视图与仰视图。**本页图**: 功能地图。**对页图**: 夜景效果图。

摩天大楼见证了大地原有的特质在城市化过程中的消解：土地的自然肌理已被边界明确的建筑肌理所取代，因而人们很容易在卫星地图上区分城市地区和非城市地区。另一方面，地平线不再重要：摩天大楼塑造的天际线定义了大地与天空的关系。

"珀尔修斯之盾"试图提供一个全新的视角来看待摩天大楼是如何改变天地间的景观的。一面巨大的正方形镜面，悬浮于既有的摩天大楼之上，上面能反射一部分的天空，下面能反射一部分的城市。

它能被放在世界上的任何城市之上。"珀尔修斯之盾"并不意味着对城市化进行美杜莎式的指控，相反它给人们一个既能身处城市之中，同时又能从远处审视这个城市的机会。在这个意义上，这个方案既是摩天大楼族群的一员，又是摩天大楼的反思者。

Skyscrapers witness the decomposition of the original characteristics of the land in the process of urbanization: The natural texture of land is replaced by the building texture with clear boundary, so it is easy to distinguish urban areas and non-urban areas in the satellite map. The horizon, on the other hand, is no longer important: The skyline shaped by the skyscraper defines the relationship between the earth and the sky.

"Perseus Shield" attempts to provide a new perspective to investigate how skyscraper changes the landscape of the universe. A huge square mirror, suspended from the existing skyscraper, will reflect a part of the sky above and a part of the city below.

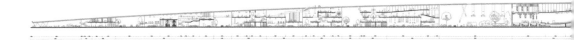

It can be placed on any city in the world. "Perseus Shield" does not mean the Medusa's allegations of urbanization. On the contrary, it gives people an opportunity to live in the city while examining the city from a distance. In this sense, the project is not only a member of the skyscraper group, but also a reflection of skyscraper.

跨页图: 剖面图。对页图: 夜景效果图。本页图: 白天效果图。

本页图:人视仰视效果图。**对页图**:
半鸟瞰效果图。

教师点评

北京的上空出现了一面镜子,反射着自己,让你从不同角度观察自己,这更像一个哲学、寓言式的作品。在空中的这一平方公里,提供了一个新的媒体建筑的空间,高高在上,脱离现实,反射真相,而又非常吸引人去互动参与。

Teacher's comments

A mirror, appearing in the sky above Beijing, reflects itself and makes you observe yourself from all angles. It is more like a philosophical and allegorical work. The one square kilometer in the sky provides a new space for the media building, surreal, high up in the air, reflecting the truth and attracting people to participate in the interaction.

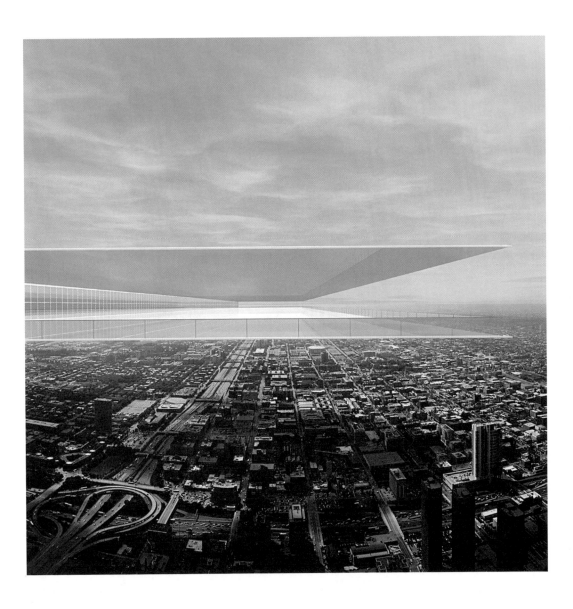

猴子的森林/
MONKEY FOREST

项目选址：CCTV 新址
项目类型：CCTV
建筑面积：50 万 m²

方案设计：盖郑
指导教师：马岩松
完成时间：2014

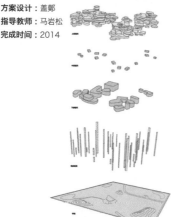

漫步在北京的CBD，遍地的高楼和繁忙的交通充斥着我的视线，很难找到一片宁静祥和的绿地稍作喘息。在这里生活的人们，头顶灰暗的天空，呼吸着烟尘和尾气；为生活承担着巨大的压力，却难有片刻的放松和休息。

相比于都市中奔波劳碌的人群，在某种意义上，猴子是幸福的，它们可以在森林中自由地奔跑、翻滚、跳跃，可以享受蔚蓝的天空，呼吸清新的空气。

本设计旨在打破当前CBD拥塞的局面，建筑像一片森林一样坐落在CBD中间。建筑可以分为上中下三部分，下层部分好比森林中的灌木丛，通过安排开放性的功能使其得以向外界开放，为在CBD工作的人们和游客带来一个可以回归自然的环境，在工作之余可以来此放松身心；中层的建筑零散错落的掩映在树丛中，主要为演播室和酒店功能；"树冠"部分则是CCTV区域，通过丰富有趣的娱乐设施和复杂多变的空间架构，为在此工作的人们创造不同寻常的工作、生活环境，让这里的生活变得多姿多彩。

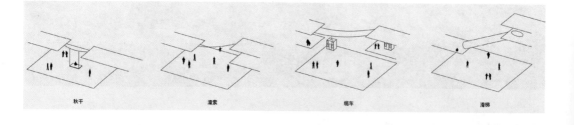

城市触媒 / CATALYZER OF CITY

项目选址：北京市朝阳区东三环中路
项目类型（功能）：电视台
建筑面积：60万 m²
用地面积：20万 m²

方案设计：金世雄
指导教师：马岩松
完成时间：2014

旧城市的问题：自说自话的建筑；重复的元素；高层之间的隔离；冰冷的机器。

本设计要带来的是：在城市中一个不同的元素；不是作为准则，而是作为促进城市空间变化的载体；打开自身才能融合环境；作为区域中心，有汇聚信息能力的生命感。

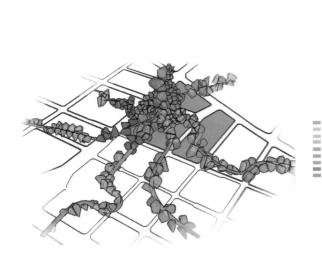

在云端/UP IN THE AIR

项目选址：北京市朝阳区光华东路
项目类型（功能）：中央电视台总部大楼
建筑面积：420 000m²
用地面积：197 000m²

方案设计：邬慧丽
指导教师：马岩松
完成时间：2014

以"云"为概念，打破传统形式摩天楼在基地上闭塞独立、且在空间上缺乏公共空间的状态，将一种异形的空间插入传统摩天楼，并在两者冲撞处形成大型的半室外公共空间，将绿化景观与休憩娱乐功能引入严肃的摩天楼形式，形成新的富于生机的空间。

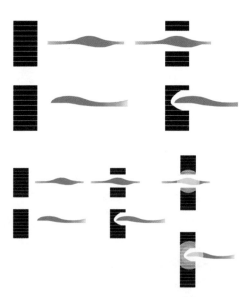

天穹花园/DOME ABOVE GARDEN

项目选址：北京市朝阳区光华路原CCTV大楼所在地
项目类型（功能）：广播电视台大楼
建筑面积：50万 m²
用地面积：6.4万 m²

方案设计：谢骞
指导教师：马岩松
完成时间：2014

在调研过程中，有两个地方给我的感受很深。一个是吴耀东老师在接受我们采访时讲的一段话，"这么大一个建筑，为什么要放在CBD？CBD那本来已经很堵了，为什么又放进去这么大一个瘤？导致CCTV的人出不去也进不来。" 另外一个是当我在实地调研的时候，发现这块地段附近没有一个可以歇脚的地方，而且在CBD附近也缺少公共空间。

基于以上两点，首先，我想把建筑提高，在建筑下方营造一个公园。而建筑横跨公园的上空，形成类似于桥或者虹的感觉。建筑里的人仿佛生活在树冠上，而在公园里的人仰望天空时，可以隐约看见一道建筑横跨天空。此外，建筑的底面可以反射公园的景象，使蓝色的天空与反射的绿影映衬在一起。功能的安排也与建筑形式相结合。

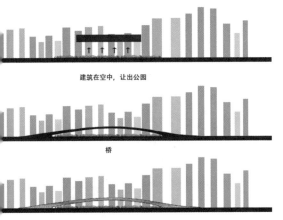

建筑在空中，让出公园

桥

结构之间形成空间

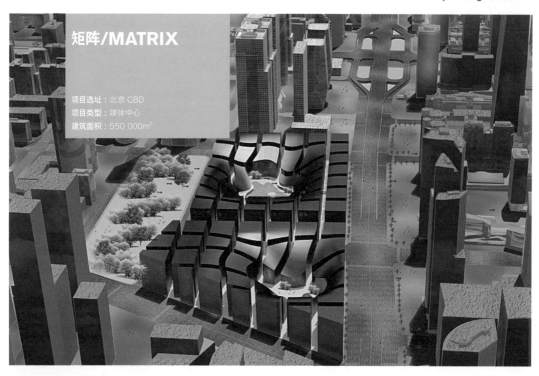

矩阵/MATRIX

项目选址：北京CBD
项目类型：媒体中心
建筑面积：550 000m²

方案设计：徐晨宇
指导教师：马岩松
完成时间：2014

这个设计一开始对地段中那些标榜权力和资本的建筑持的是不妥协的态度，所以采用了阵列低矮方体的方式，希望以此产生与那些高耸建筑相当的力量。同时，通过这个矩阵把当代所批判的现代主义"钢筋混凝土的丛林"推向极致，却发现它与周围的建筑相比显得更加自律，这未尝不是一种讽刺。更进一步，在讽刺之后更需要的是积极的态度。这个矩阵的内部不是像外表那样冰冷，是一种温暖的氛围，方体变得柔软，甚至融化，象征着从内部瓦解这种极端的体制，正如北京所受体制带来的限制逐渐消解。内部产生出柔和的曲线、开放的界面和宽阔的城市空间，用这种方式面对民众，也表现CCTV与民众的互动与关注人民大众的态度。。

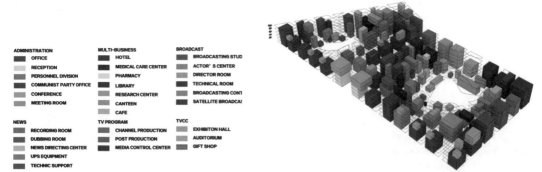

在树下/UNDER THE TREE

项目选址：北京商务中心区
项目类型（功能）：中央电视台新台址
建筑面积：100万m²
用地面积：187 000m²

方案设计：张裕翔
指导教师：马岩松
完成时间：2014

树
建筑以城市为土壤
苍劲的枝干上擎天际
谦虚的枝叶倒挂下来
以有限的绿荫庇护无限
的大地

信息
媒体是信息时代的灵魂
是捉摸不定的七彩火种
我将她悬在空中
光芒辉映更广阔的城市

你中有我
是建筑亦是城市
正反天际线相互排斥
互相吸引
释放出耐人寻味的场域

[模型照片]

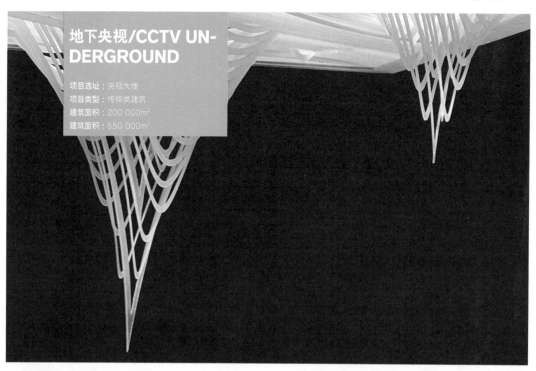

地下央视/CCTV UNDERGROUND

项目选址：央视大楼
项目类型：传媒类建筑
建筑面积：200 000m²
建筑面积：550 000m²

方案设计：赵健程
指导教师：马岩松
完成时间：2014

设计的初衷在于为日益拥挤的城市营造一方空间。调研初期，我把目光放到了20世纪五六十年代的美国。此时资本社会与工业革命的成就广遭质疑，艺术家们用大地作为载体，呼唤野性与和平。Agnes在曼哈顿最繁华的地界种植了两亩小麦，在"冲突"中警示人们，不要走得太远而忘记为什么出发。

我的CCTV亦是如此。

老子云："玄牝之门，是谓天地根。"深埋地下的锥体如女性生殖器一般，神圣而又神秘，在讥讽着摩天大楼的同时，为人们开启另一片属于地下的世界，然后还天地以光明。

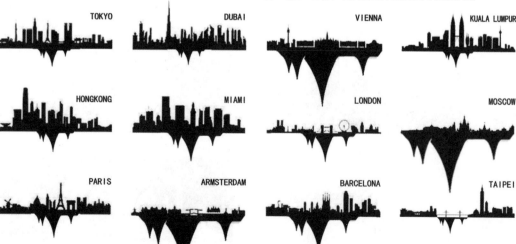

城市沙堡/CITY SAND CASTLE

项目选址：北京朝阳 CBD 现 CCTV 地址
项目类型：CCTV 大楼
场地面积：19.7 万 m²

方案设计：周辰
指导教师：马岩松
完成时间：2014

玻璃、钢铁、距离
摩天楼以一种高不可攀的视角俯视这个城市
沙子、粗糙、温暖
也许我们要让城市丛林里多一些亲近与暖意
精致、光滑、模数化
现代主义拒绝一切不可控不确定的因素
随性、自然、想象
我们可以摆脱条条框框的束缚去解放自己
高度、效率、垂直电梯
现代人际交流似乎越发简短和有目的性
洞穴、连通、水平
交流应当回归到原始的那种多样和未知的可能性
生活只是一场每日照例上演的舞台剧
还是我们永远能在一个不经意的角落发现未知的自己
城市究竟是一个由程序掌控的冰冷的matrix
还是茫茫大海边人类创造的粗糙与亲近的一颗沙砾

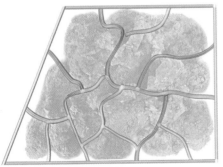

平面上的连接，开放式的场地

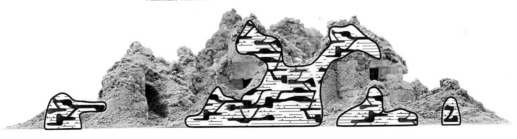

魔方/CUBE

项目选址：东三环 CCTV 现址
项目类型：办公综合体
建筑面积：55 000m²
建筑面积：19 000m²

方案设计：朱吴孟健
指导教师：马岩松
完成时间：2014

CCTV作为一个中国特色的国家媒体，其警戒保卫的封闭特性导致了无论放在哪里，对周围的城市环境都有比较大的影响。这是建国以来根植于北京办公建筑的一种模式，"大院模式"。如果这种模式无法改变，那么可以让自组织的"大院"占用的城市地面空间尽可能小，把地面作为公园还给市民。

CCTV与城市的关系更像是一种天外来客般的文明碰撞。库哈斯的中标提醒了我们CCTV作为甲方，并不想消隐在城市中，而是要在城市中创造一种奇异的景观。

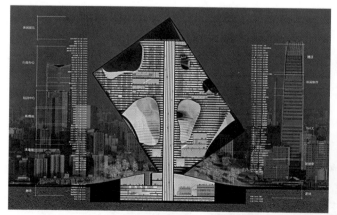

天空之树 /THE SKY TREE

项目选址：北京市朝阳区光华路原 CCTV 大楼所在地
项目类型（功能）：广播电视台大楼
建筑面积：100 万 m^2
用地面积：1 万 m^2

方案设计：邹尚武
指导教师：马岩松
完成时间：2014

当今社会各个方面都对摩天楼进行着批判，然而各个时期依然没有停止对摩天楼的修建，而且越来越高，因为这象征着政治、经济、权力。楼的高度也从19世纪的保险公司大楼的50m到现在迪拜塔的800m。人类的视野不断地提升，我们所看见的东西也越来越有所不同——在世界上任何地方的高楼上都能看见如织的人群面对着前所未见的景色而发出阵阵惊呼。因为无论在何时何地，人对于未知世界的好奇以及探索是永无止尽的。那么我们不得不思考，为了满足人类的追求与欲望，摩天楼的极限在哪，我们又能到达怎样的高度呢？摩天楼从来没有摆脱与重力的对抗，而在科技技术飞速发展的未来，摩天楼能不能摆脱重力的束缚？在这样的思考背景下，也许这个永远不能实现的方案能够打破人类的常规思维，将我们的视野延伸到太空……

要达到如此高度还会因为空气发生倾斜，那么建筑的形象能不能像竹节一样呢？充满韧性随风摇摆。

理想

华黎

迹·建筑事务所 创始人、主持建筑师

自宅

华黎
迹·建筑事务所
创始人、主持建筑师

教育背景
1989年 – 1994年
清华大学建筑学院建筑学 学士
1994年 – 1997年
清华大学建筑学院建筑学 硕士
1997年 – 1999年
美国耶鲁大学建筑学院建筑学 硕士

工作经历
1999年 – 2001年
纽约Westfourth Architecture P.c.建筑事务所 建筑师
2001年 – 2002年
纽约HB+FR& Associates 建筑事务所 建筑师
2003年 – 2008年
普筑设计事务所 合伙人
2008年至今
迹·建筑事务所 创始人、主持建筑师

主要论著
《建筑界丛书第二辑——起点与重力》[M]. 中国建筑工业出版社，2015/11
《四分院札记》[J]. 建筑学报，2015/11
《近思远观——华黎谈作为基地的地景》[J]. 建筑师，2015/10
《建筑之轻》[J]. 建筑学报，2015/07
《回归本体的建造》[J]. T+A 时代建筑，2014/05
《华黎访谈》[J]. art4d（泰国），2014/01
《北京迹·建筑事务所：TAO华黎用当地建造技术在偏远地区建造》[J]. MARK（荷兰），2013/12
《2013阿卡汗建筑奖专辑 高黎贡手工造纸博物馆》[J]. WA 世界建筑，2013/11
华黎，《Tracing the Roots, Museum of Handicraft Paper》，收录于《Homecoming: Contextualizing, Materializing and Practicing the Rural in China 》[M]. 2013/10
《此时此地——华黎访谈 在地：华黎的建筑之道和他的建造痕迹》[J]. 世界建筑导报，2013/10
《起点与重力》[J]. 新观察，2013/06
《微型城市——四川德阳孝泉民族小学灾后重建》[J]. 建筑学报，2011/07
《建造的痕迹，云南高黎贡手工造纸博物馆》[J]. 建筑学报，2011/06
《 微缩之城——四川孝泉民族小学灾后重建 》[J]. domus，2011/05
《Exemplary Metamorphosis》[J]. MD（德国），2009/12

设计获奖
2013.10 – 孝泉民族小学获2013亚洲建协建筑奖
2013.10 – 华黎获得北京国际设计周2013 D21中国建筑设计青年建筑师奖
2013.05 – 高黎贡手工造纸博物馆入围2013阿卡汗国际建筑奖短名单
2012.12 – TAO入选美国建筑实录杂志评选的全球设计先锋
2012.12 – 华黎获得第三届中国建筑传媒奖青年建筑师奖
2012.12-2012 – WA中国建筑奖优胜奖高黎贡手工造纸博物馆
2012.12 – Design for Asia (DFA) 孝泉民族小学获得2012香港亚洲设计奖荣誉奖
2012.07 – 美国《建筑实录》杂志"好设计创造好效益"中国奖"最佳公共建筑奖"四川孝泉民族小学，高黎贡手工造纸博物馆

代表作品
三里屯甜品咖啡店(图1)、岩景茶室(图2)、四分院(图3)、林建筑(图4)、高黎贡手工造纸博物馆(图5、图6)、武夷山竹 筱育制场(图7、图8)、四川孝泉民族小学(图9)

理想自宅

三年级建筑设计（6）设计任务书

指导教师: 华黎
助理教师: 李若星

在这8周里，你将为自己或你非常熟悉的人设计一栋自宅。你可以根据你自己的兴趣假定房子主人的身份（例如:建筑师、设计师、音乐人、电影导演、艺术家、作家、或其他）。首先，你需要谈出对房子主人生活的理解，这可以从你自己的感受和体会出发，或是曾经非常触动你的文学作品或电影中的场景出发），自己拟定主人对空间的需要（例如功能空间、家庭成员、事件等）并进行阐释。这类似于每一个人都讲述一个有意思的故事，然后大家一起讨论。

"So you look into the nature, and then you are confronted with the program. Look at the nature of it, and you see in the program that you want.... The first thing that is done is the rewriting of the program." ——Louis Kahn

之后，自己拟定一个基地，这个基地可以是你去过的真实的某个地方，符合你对生活的设想，也可以是你自己虚拟的基地，带有某些你需要的条件（例如地形、景观、朝向、气候、氛围），也可以完全抽象没有具体条件。但每个人必须要阐释自己对基地的感受、理解和想法，与大家分享。

在此基础上，你将开展设计。记住: 在这个设计中，建筑始终都应被理解为场所，建筑实际上只是一个界面，人与自然的界面，以及人与人的界面，因此一定不要只看到建筑本身作为一个物质体，而更应该关注它所能承载的生活。最终对于建筑的表达同时也是讲述建筑内部的人的故事。

时间安排

第一周，阅读、空间故事的讨论
第二周，拟定空间需求、拟定基地、开始初步构思
第三周，初步设计想法,草图、工作模型
第四周，设计的相互评价讨论
第五周，中评（外请老师）
第六、七周，深化
第八周，终评（外请老师）

设计要求: 建筑面积不超过 500 m^2，其他自定。

表达要求

1:100 模型(含基地);

1:30 模型;应有家具、陈设、人物、材料的表达模型照片,表达空间、光线、材质、氛围构思草图

平面图 1:50

剖面图 1:50

重点部位墙身构造图 1:20(选择性)

推荐阅读:卡尔维诺——《看不见的城市》, 卡尔维诺——《月亮的距离》, 博尔赫斯——《迷宫》(Labyrinth),卡夫卡——《城堡》,路易·康——《对房间的论述》

Beatriz Colomina, The Split Wall: Domestic Voyeurism;藤本壮介——《建筑诞生的时刻》,路易·康阿兰·德波顿——《幸福的建筑》

推荐电影

Luis Bunuel: The Discreet Charm of the Bourgeoisie《中产阶级的谨慎魅力》,1972

王家卫《重庆森林》,1994

Jean-Luc Godard, Contempt《蔑视》,1963

Michelangelo Antonioni, BLOW-UP《放大》,1966

Peter Weir, The Truman Show《楚门的世界》,1998

推荐案例分析

Raimund Abraham, House without rooms Adolf Loos, Muller House

Peter Zumthor, Chapel in the field, Koln Louis Kahn, Exeter Library

Jorn Utzon, House in Mallorca

Carlo Scarpa, Canova Museum

临崖之宅/
HOUSE ON THE CLIFF

项目选址：一座设想的悬崖（基于爱尔兰的莫赫悬崖）
项目类型：自宅
建筑面积：408m²

方案设计：张文昭
指导教师：华黎
完成时间：2014

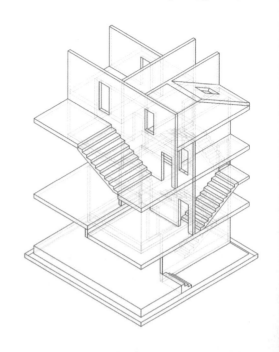

这个设计首先是一段发现自我的内心旅程。"理想自宅"意味着一个理想的人格和一种理想的生活方式。对于我来说，这是像法国哲学家蒙田那样的：一种相对Stoic的生活，于自然中追求精神塑造和内心宁静，隐居的同时对生活保有质朴而亲近的态度。在这座崖上之宅中，房间以灵活的标高在垂直方向上围绕楼梯井组织，人在其中会经历盘旋向上的序列化的体验。从底层的厨房，到会客室，再到顶部的书房和爬梯子才能到达的冥想阁楼，自下而上是从mundane的日常生活到更加精神化和个人化的spiritual life的逐渐过渡。这是一个从相对公共到更加个人化的转换，也是一个精神上升的隐喻。而光线和材质在每一个场景中都成为了氛围塑造的主要因素，引导和提示着人在空间中的活动。

Open Design Studio 2014

开篇：主要效果图。**对页上图**：轴测图。**对页下图**：1:50模型。**本页下图**：结构轴测图。**本页右图**：1:200模型。

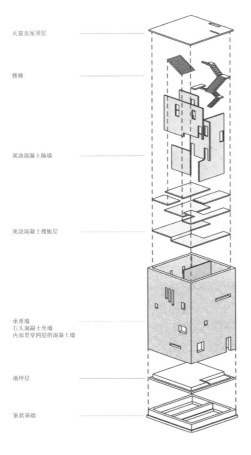

天窗及屋顶层

楼梯

现浇混凝土隔墙

现浇混凝土楼板层

承重墙
石头混凝土外墙
内部贯穿四层的混凝土墙

地坪层

条状基础

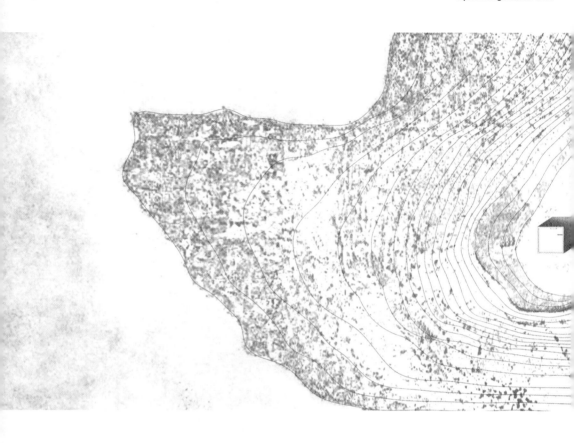

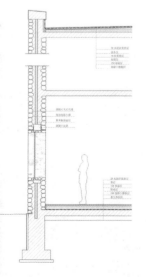

This design is first an inner journey of self-discovery. "Ideal House" means an ideal personality and an ideal way of life. As far as I'm concerned, it means a way of life with respect to Stoic, just like the life lived by Montaigne, a French philosopher, who pursues spiritual shaping and inner peace in the nature, and keeps simple and close attitude towards life while living in seclusion. In the house on the cliff, the rooms are arranged around the staircase in the vertical direction with flexible elevation. People in the rooms will feel a spiral upward sequence of experience. From the bottom kitchen, to the reception room and to the top study room and the meditation attic that can be reached by climbing a ladder, the bottom-up structure is a gradual transition from the mundane daily life to the more spiritual and individualized spiritual life, and from a relatively public to a more personal space. It's also a metaphor of spiritual rise. The light and material in each scene become main factors of atmosphere shaping, which guides and reminds of the activities of people in space.

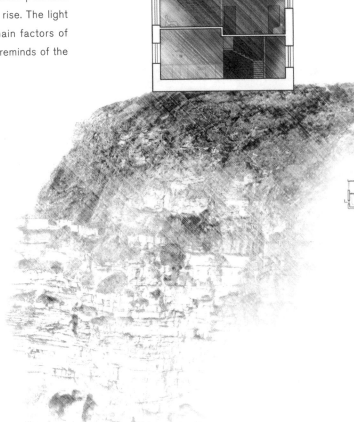

对页上图：总平面图。**对页下图**：墙身大样。**本页右图**：剖面。

教师点评

这是一个克制的设计,因其克制所以专注,因其专注所以有力。对空间的想象从感性出发,空间组织则因循一个清晰的逻辑,具有内省特征的空间序列围绕一个楼梯井盘旋而上逐渐展开,通过窗洞口形成与外部世界间歇性的碰撞,同时具有从物质生活逐渐上升到精神生活的隐喻。对空间及窗洞口比例的微妙拿捏、对自然光审慎的捕捉、以及对材料的运用,共同塑造了具有诗意和静谧氛围的空间。

Teacher's comments

It is a design of restraint. The restraint results in concentration, and the concentration contributes to powerful design. From sensibility in the aspect of imagination of space, the spatial organization follows a clear logic, and spatial series with introspective characteristics spirals up and is gradually unfolded along a stairwell. It not only forms intermittent collision with the outer world through the cave of window, but also creates the metaphor of gradually rising from material life to spiritual life. Subtle balance of proportion of the space and cave of window, careful capture of natural light and application of materials jointly create a space with poetic thought and quiet atmosphere.

上图（从左到右）：室内核心透视—书房。楼梯间。室内透视—起居室。
对页下图：楼梯间。本页下图：室内透视—厨房。

双重人格宅 / HOUSE FOR TWO PERSONALITY

项目选址：内心
项目类型：自宅（居住）
建筑面积：500m²
用地面积：150m²

方案设计：程正雨
指导教师：华黎
完成时间：2014

开篇: 透视图。**本页左图**: 主方案图。
对页: 分析图。

本方案是一个双重人格分裂者的自宅。在整个建筑中有两种属性的空间,分别对应两个人格。这两种空间有各自的连续性,同时也有突变性,即可以从一种空间转换到另一种空间。在两种性格迥异的空间转化过程也隐喻着人格的转换。
两种空间各自通过空间的尺度、颜色、材料、光影等的区别来营造两种完全不同的氛围,一个阴暗冷酷,认真严苛,逻辑缜密;一个阳光开朗,感性自由,想象丰富。这两种类型的空间缠绕式上升,表达出两种人格相互影响的状态。同时在外面的状态又同属一种颜色,合成整体,来体现在一个躯体当中两个人格存在的状态。

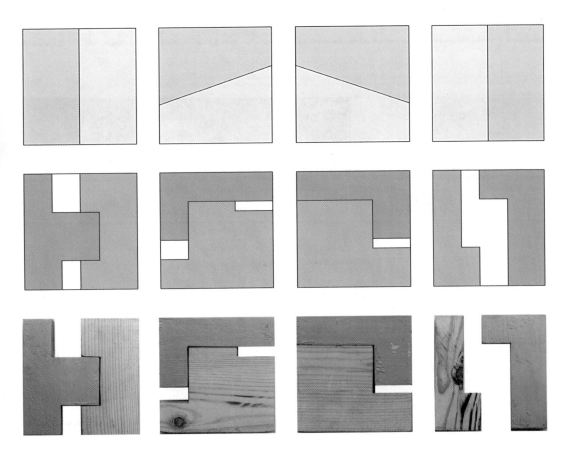

This scheme is the house of a person with double split personality. In the whole building, there are two types of spaces with two attributes, respectively corresponding to the two personalities. These two kinds of spaces have their own continuity, as well as the mutability, which can be converted from one space to another space. The space transformation process for two different types of personalities also correspond to the conversion of personality.

Two spaces respectively create two completely different atmospheres through the different arrangement of scale, color, material and lighting; one space is dark

and cold, serious and harsh, rigorous and logic; the other space is sunny and cheerful, emotional and free, and full of imagination. These two types of spaces wind up, expressing the status that the two personalities are affected mutually. At the same time, the outside states contain only one color, which is integrated into the whole, reflecting the status of two personalities in one body.

对页上图：平面与剖面图。**本页上图**：立面图。**跨页图**：剖面图模型。

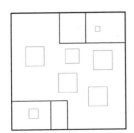 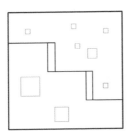 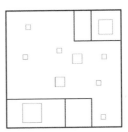 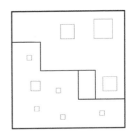

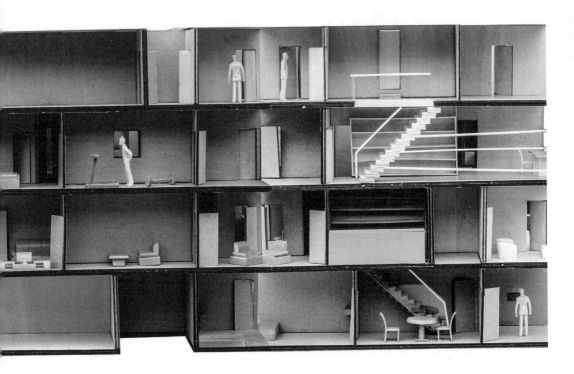

教师点评

设计从人格分裂这一概念推演而出并准确找到了空间组织的清晰逻辑,这一出发点本身颇具巧思,而设计过程则因清晰而顺畅。对应于双重人格,两种特质的内部空间以双螺旋的方式相互缠绕,在一个纯粹的立方体内完成了组织与转换关系,使其免于无关形式的困扰,因而具有更为抽象的原型意义。

Teacher's comments

The design derives from the concept of split personality, and helps find clear logic of spatial organization accurately. This starting point is provided with ingenuity, and the

design process is clear and smooth. Corresponding to dual personality, inner spaces of the two traits spiral each other in the way of double helix, and complete organization and transformation relation in a pure cube, so as to avoid perplexity of irrelevant forms. So it is provided with more abstract prototype meaning.

对页上图：相遇。跨页图：平静。本页右图（从上至下）：黑暗、光明、对望、再会。

三口之家
HOUSE FOR FAMILY OF THREE

项目选址：故乡河边一片狭长的树林里
项目类型：自宅
建筑面积：306m²
用地面积：203m²

方案设计：谢湘雅
指导教师：华黎
完成时间：2014

这个住宅是为一个三口之家设计的，其中也体现着我自己的生活经历和对家庭生活的理解。在这个家庭中，每个家庭成员都有着极强的家庭归属感。他们并不热衷社交，认为家庭是栖息地，也是幸福的来源，是自我认同和自身价值的根基。在这个住宅中，我希望营造的家庭生活是这样的：房间之间并不是完全封闭，而是存在着视线的贯通、声音的穿透和气味的扩散，家庭成员可以时刻感受到彼此。通过墙面洞口和高差，组织平面和剖面，营造空间之间的视线交流。当然家庭中的每个人依然需要有隐私、自省和孤独，这个住宅中也设计了一些空间层次，保证了每个家庭成员的个体需求。

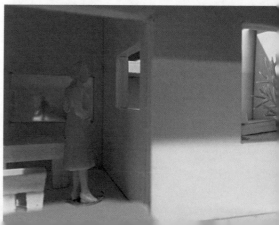

开篇:主方案图。**本页上图**:地段模型。**跨页图**:对视场景。

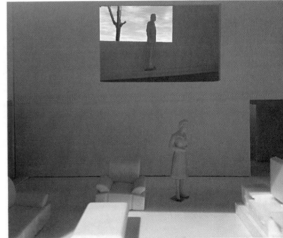

This house is designed for a family of three, which also reflects my own life experience and understanding of family life. In this family, each member has a strong sense of belonging to the family. They are not keen on social networking, thinking that the family is habitat, as well as the source of happiness, which is the foundation of self-identity and personal values. In the house, I want to create a family life as follows: The space among rooms is not completely closed, but there is direct sight, penetrated voice and diffusion of smell and the family members can always sense each other. Through the opening on the wall and the height difference, plan and section are organized to create the communication by sight among the spaces. Of course, everyone in the family still needs to have privacy, introspection and solitude. The house is also designed with the space level to ensure the individual needs of each member of the family is satisfied.

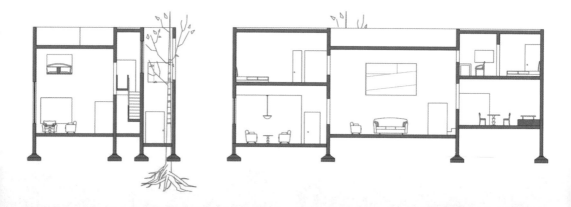

对页上图：剖面图。**对页下图**：模型外壳。**本页上图**：平面图。**本页下图**：模型内部。

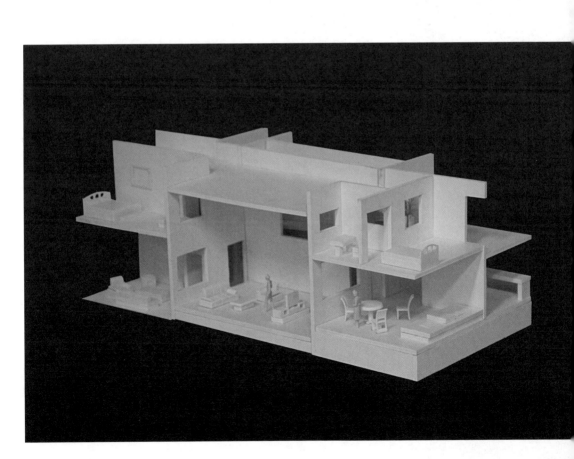

教师点评

这个设计从一个三口家庭中个人空间的内在需求出发来塑造单个空间,而不同空间之间的多重关系又塑造了家庭内部故事的结构,单个空间的形式遵从于其自身的逻辑,而空间序列组织上则通过比例、尺度、明暗等节奏变化完成了整个建筑的空间叙事。墙作为建筑中最基本的元素来限定空间,空间之间的联系则由洞口来完成,这也使得房子内部的空间产生了视线上的多层关系。形式上朴素而清晰的语言使建筑获得一种整体性。

Teacher's comments

From inner demand of personal space in a family of three, this design shapes a single space. Multiple relations between different spaces also shape the structure of stories within a family. The form of a single space follows its own logic, while the organization of spatial series completes spatial narrative of the whole building through scale, dimension, brightness and darkness, and other rhythm changes. Walls limit the spaces as the most basic element of a building. Relations between spaces are completed by opening, which also makes inner space of the house generate multilayer relationship from line of sight. Simple and clear language in form makes the building obtain a kind of integrity.

对页图:主卧—餐厅。本页上图:书房。本页下图:主卧—望向河岸。

巴学园/BUS STUDY GARDEN

项目选址：虚构
项目类型：理想自宅
建筑面积：200m²
用地面积：500m²

方案设计：陈凌亚
指导教师：华黎
完成时间：2014

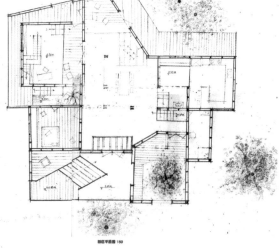

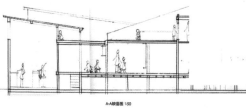

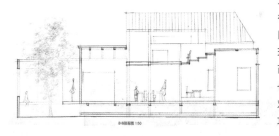

这是一个由书本出发的方案设计，选取的书籍是日本作家黑柳彻子的《窗边的小豆豆》，书中描写了作者童年时代的巴学园，一个建立在电车之上的学校。书中小孩子对世界对空间的理解方式让我觉得十分有趣，所有的构件都没有特定的功能而是对应着当时的使用需求而发生改变。因此，在设计之中我也尝试着把书架和楼梯、屋顶和滑梯等各种元素组合起来，对应各种使用需求，设计的过程就像搭积木一样，充满偶然和巧合这些正是本设计要达到的效果。

迟子建自宅/CHI ZI-JIAN HOUSE

项目选址：高层住宅楼顶
项目类型：住宅
建筑面积：190.5m²
用地面积：223.5m²

方案设计：范孟辰
指导教师：华黎
完成时间：2014

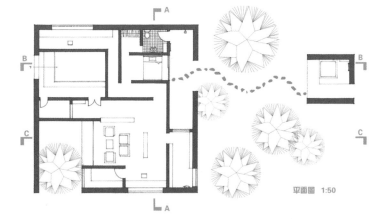

平面图 1:50

本设计是为女作家迟子建设计的自宅，迟子建认为现代技术的发展使得人们变得麻木，空调和暖气让人们无法真正的接触自然。设计在外部形成一座堡垒，主要想要表现作家面向自我、内省的特质，以及她的思想被现代社会和现代生活孤立的局面。内部塑造的各个空间使得作家在自宅中的各处可以触到自然的气息。

在用餐时眺望整个城市的风光

修行宅/HOUSE FOR MEDITATION

项目选址：中国某南方城市南方城市
项目类型：住宅
建筑面积：296m²
占地面积：144m²

方案设计：高祺
指导教师：华黎
完成时间：2014

这是一座位于城市地段的独居住宅，居住者为现代修行的人。

从二层入口选择向上进入开敞的生活空间或向下至内向的精神空间，两类空间通过体量间隙进行交流而又保持各自独立。

内院所栽的树在多方向与人保持或远观或亲近的关系，为居住者建立与自然、内心和信仰的连接提供可能。

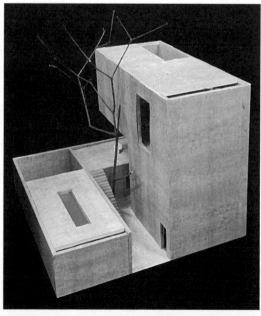

墙宅/HOUSE IN THE WALLS

项目选址：后海民居
项目类型：小型住宅
建筑面积：210m²
用地面积：250m²

方案设计：郭金
指导教师：华黎
完成时间：2014

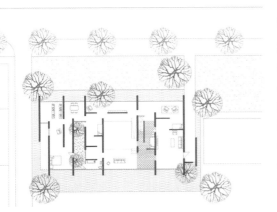

外面的世界很复杂，所以别再让家变得看似复杂。对我而言，"理想自宅"是对一种生活状态、生活方式的向往和实践——简单逻辑下的自由生活。

所以这所长在后海边上的房子，回归建筑最基本的元素，以墙为引导，让视线里的后海无遮无挡。

伸出屋顶的砖墙，承载着重量的混凝土墙间隔排列，成为空间序列中两张有趣的脸。缝隙从中生长：墙间的缝隙激活了空间；墙与楼板间的缝隙点亮了空间。满足结构要求的同时，引入了光和自然。

像在展览馆一样条条列列的生活场景中，有庭院，有水池，而让这些丰富的空间生动地联结起来的是这些隐约缝隙做出的指引。

音乐之家/A HOUSE FOR H.O.T

项目选址：韩国首尔江南区
项目类型（功能）：别墅
建筑面积：396m²
用地面积：900m²

方案设计：刘涵
指导教师：华黎
完成时间：2014

这是一个给韩国过气偶像音乐人的基地，在某种意义上来说他们代表了那一代的偶像。对于他们来说，生活大致可以分成五个部分：交流聚会、音乐创作、作品练习、回忆过去以及自我反省。在这个基地中，五个元素根据不同的需求而转换变为五个空间。

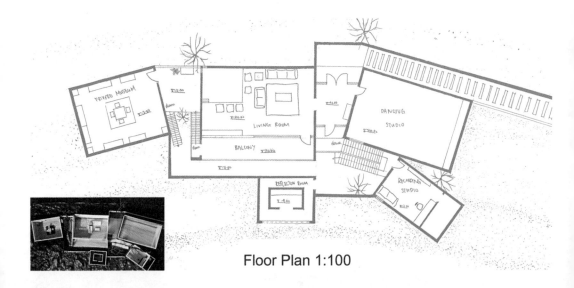

Floor Plan 1:100

流浪者的家/HOUSE OF THE NOMAD

项目选址：北纬40°的某地
项目类型（功能）：理想自宅
建筑面积：640m²
用地面积：360m²

方案设计：刘奕君
指导教师：华黎
完成时间：2014

房子的主人自我定义为一个流浪者，他的职业是设计师。这个建筑是他精神世界的投射和隐喻。垂直的圆筒既是建筑的精神性空间，也是结构支撑。不同朝向的水平盒子容纳了主人的日常生活，从上到下依次是工作空间、睡眠空间、起居空间；盒子的朝向由主人的作息时间，一天中各个时间段对阳光的不同需求和所在地区纬度的太阳轨迹决定（此处设定为北纬40°）。

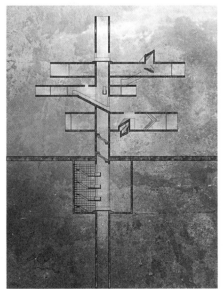

天街庐/HOUSE OF ENCOUNTER

项目选址：重庆市长寿区三道拐老街
项目类型：自宅
建筑面积：215m²
用地面积：260m²

方案设计：卢清新
指导教师：华黎
完成时间：2014

雾失楼台，月迷津渡，而那缥缈的空中，便是我美丽的街市，窄窄的街旁，便是我理想的自宅。老木头的沉郁味道，直棂窗的曼妙疏影，麻将馆的布幔轻摇，还有惝恍中，不期而遇的故事，一切都发生在，这天街上的小房子中。

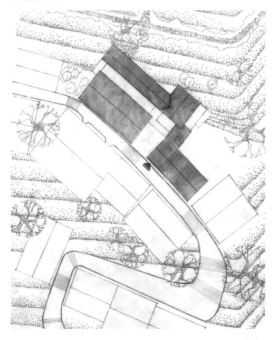

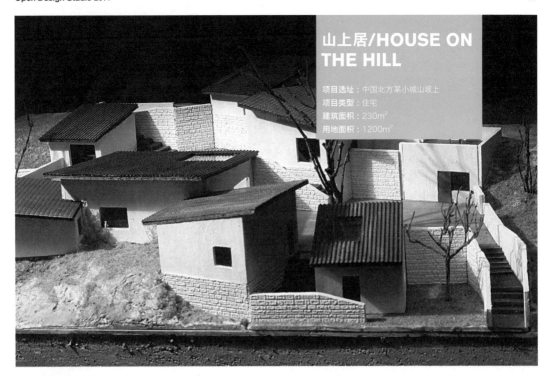

山上居/HOUSE ON THE HILL

项目选址：中国北方某小城山坡上
项目类型：住宅
建筑面积：230m²
用地面积：1200m²

方案设计：王梦熠
指导教师：华黎
完成时间：2014

童年，我的暑假在河边山坡上的外婆家度过，那是一个温暖的小村庄——在悬崖边冒险、在屋顶上行走、穿梭于屋角与巷弄。如今它已经消失，小时候陪伴我的亲人也难得见面。爱做针线活的外婆，喜欢园艺的姨妈，和我一起玩耍的表姐，我思念的亲人们，一起回到回忆之地吧。

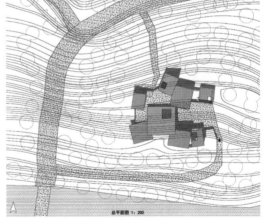

总平面图 1:200

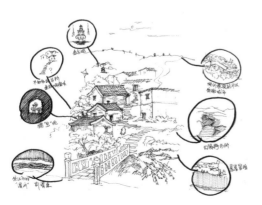

水上村落 / MICRO VILLAGE ABOVE WATER

项目选址：云南大理洱源西湖洱海
项目类型（功能）：自宅
建筑面积：450m²
用地面积：600m²

【初霁】

方案设计：朱玉风
指导教师：华黎
完成时间：2014

【断桥】

【学生】

【月夜】

【路途】

■ 理想自宅·结构研究
Ideal House - Structure Study

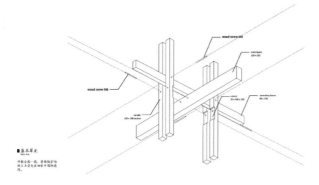

■ 基本单元

云南大理洱源西湖中的小岛，热爱乡土的知识分子一家四口的自宅。设计关乎传统村落记忆——灶上熏火腿，美人靠上听雨声，孩子在老屋袒露的木结构上的攀爬……亦关乎场地的处理——高差的应用，树的位置，湖的视野，干湿交替水位涨落给地段带来的机遇和挑战……

关乎结构的沿袭和创新，材质的运用，氛围的捕捉……

生活的

李兴钢

中国建筑设计院总建筑师
李兴钢建筑工作室主持人

李兴钢
中国建筑设计院总建筑师
李兴钢建筑工作室主持人

教育背景
1987年－1991年
天津大学建筑学院 建筑学学士
2006年－2012年
天津大学建筑学院 工学博士

工作经历
1991年至今
中国建筑设计院有限公司（原建设部建筑设计院）
历任助理建筑师、建筑师、高级建筑师、教授级高级建筑师，院副总建筑师
现任院总建筑师
李兴钢建筑设计工作室主持建筑师

主要论著
《静谧与喧嚣》建筑界丛书第二辑[M].中国建筑工业出版社，2015/11
《李兴钢：胜景几何》[M].城市/环境/设计（UED），中国建筑工业出版社，2014/01
《Li Xinggang: Geometry and Sheng Jing》[M].Urban/Environment/Design（UED），中国建筑工业出版社，2014/04,001
《李兴钢》当代建筑师系列[M].中国建筑工业出版社，2012/08
《瞬时桃花源》[J].建筑学报，2015/11
《静谧与喧嚣》[J].建筑学报，2015/11
《留树作庭随遇而安折顶拟山会心不远——记绩溪博物馆》[J].建筑学报，2014/02
《探求建筑形式、结构与空间的同一性——海南国际会展中心设计手记》[J].建筑学报，2012/07
《虚像、现实与灾难体验——建川文革镜鉴博物馆暨汶川地震纪念馆设计》[J].建筑学报，2010/11
《表皮与空间-北京复兴路乙59-1号改造》[J].建筑学报，2008/12
《国家体育场设计》[J].建筑学报，2008/08
《由国家体育场的设计看建筑向本原回归的倾向》[J].世界建筑，2008/06

设计获奖
国家体育场
2008 – "全国优秀工程勘察设计金质奖"
2009 – "中国建筑学会建筑创作大奖"
2008 – "第八届中国土木工程詹天佑奖"
2009 – "国际奥委会和国际体育与休闲建筑协会金奖"
复兴路乙59-1号改造
2011 – "全国优秀工程勘察设计银质奖"
2009 – "全国工程勘察设计行业奖建筑工程一等奖"
2009 – "北京市优秀工程设计二等奖"
绩溪博物馆
2015 – "全国优秀工程勘察设计行业奖建筑工程公建一等奖"
2015 – "第十八届北京市优秀工程设计一等奖"
2014 – "WAACA中国建筑奖城市贡献奖佳作奖"
海南国际会展中心
2013 – "全国优秀工程勘察设计行业奖建筑工程公建一等奖"
2013 – "中国建筑设计奖（建筑创作）金奖"
2014 – "十二届中国土木工程詹天佑奖"
2013 – "北京市第十七届优秀工程设计一等奖"
华润西柏坡希望小镇
2015 – "住房城乡建设部第一批田园建筑优秀作品一等奖"
建川镜鉴博物馆暨汶川地震纪念馆
2011 – "全国工程勘察设计行业奖建筑工程三等奖"
2011 – "北京市第十五届优秀工程设计一等奖"
2011 – "第六届中国建筑学会建筑创作奖佳作奖"
北京地铁昌平线西二旗站
2012 – "北京市第十六届优秀工程设计三等奖"
2011 – "International Achievement Awards Outstanding A"
兴涛接待展示中心
2002 – "英国世界建筑奖提名奖"

代表作品
绩溪博物馆（安徽）（图1）、复兴路乙59-1号改造（北京）（图2）、建川镜鉴博物馆暨汶川地震纪念馆（四川）（图3）、元上都遗址工作站（内蒙古）（图4）、海南国际会展中心（海南）（图5）、纸砖房（威尼斯）（图6）等

生活的舞台

三年级建筑设计（6）设计任务书

指导教师：李兴钢
助理教师：易灵洁 张玉婷

课程描述
建筑源于生活。本设计课程要求学生从身边熟悉的场地、环境、背景着手，自行合理设定人物、生活内容及使用功能，再从具体的人和生活出发，来逐步生成一个如同"生活的舞台"的建筑设计。使学生在这一过程中时刻意识到人、具体的生活和情感与建筑和自然的密切关联。同时通过书写文字、绘制草图、制作模型等表达方式，强调建筑设计过程中对心、脑、手合一的尝试、训练和对建筑尺度、建筑营造的体验和把握，以及建筑教育中"师带徒"方式意会言传的独特性。并且采用了团队式的组织、讨论、分享与个人独立工作相结合的方式，使学生体验建筑师职业工作的特点。

课程目标
目标1：学习掌握场地环境调研及分析——周边建筑物和环境、现存构筑物、植被和植物、地形、交通条件、文化背景、气候条件等。
目标2：人物设置和功能自拟，学习由具体的生活出发设计场景和建筑，从而使建筑对于使用者的生活产生价值和意义。
目标3：学习在建筑设计中将建筑的空间和自然进行密切关联，使之和使用者发生身体和情感的互动。
目标4：学习以草图、模型、图纸表达的方式逐步将头脑中的抽象意图和概念转化为具体的建筑设计。
目标5：学习建筑设计中的结构、材料、构造等建造意识和能力。
目标6：学习团队分工合作、分享建议、共享成果与个人独立工作相结合的建筑师执业工作方式。

建筑场地
清华建筑馆南建筑设计院西大草坪东的空地，约30m x 100m。

使用功能
自拟，但必须是特定真实环境及背景下合理及有说服力的设定。需包含生活中的居住、学习/工作、游憩、聚会/会议四种内容所需空间，应有大、中、小及室外开放空间。

建筑面积
1000~1500m²。

开放空间
场地内自然空地需保留 50%以上，并应与建筑室内空间建立密切的互动关系。

生活人物
8个主要人物及其他群众，8位主角需有名姓，并分别设置他们的国籍、文化背景、性情、职业和日常生活内容。最好是以自己熟悉的人为原型。8个人物之间有生活上或强或弱的关联。

表达方式
文字，不同比例草图，1/500、1/200、1/100、1/50 模型、正式图纸(1/500 总平面图、1/100平立剖面图、1/50详图及其他必要文字说明、分析图等)。所有过程记录表达也作为最终成果的一部分。

设计周期
8 周。

问道——古剑奇谭游戏/人生体验中心
ASKTAO

项目选址：清华大学建筑学院南侧草坪
项目类型：综合活动中心
建筑面积：1564m²
用地面积：400m²　绿化面积：2160m²
容积率：0.39m²　绿化率：0.54m²

方案设计：刘楚婷
指导教师：李兴钢
完成时间：2014

开篇：主方案图。**本页图**（从上至下）：总平面图。南立面图。剖面图。对页：区位分析图。下页：功能、流线图。

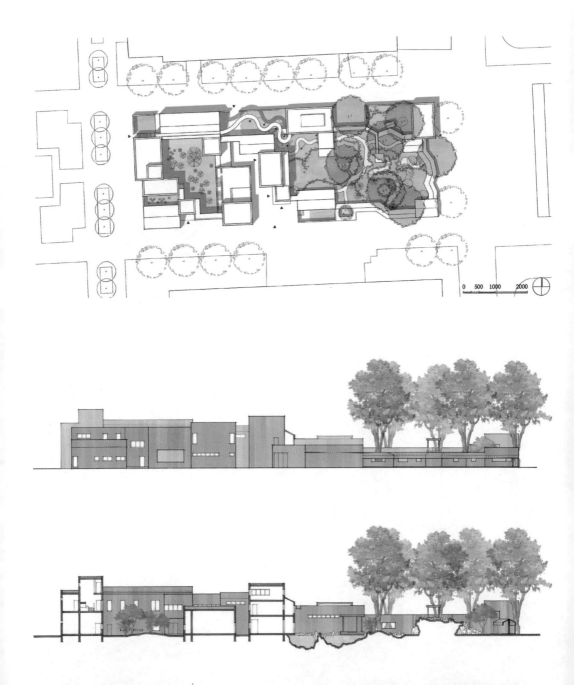

清华园前身曾为皇家苑囿，而今虽依旧毗邻圆明遗址、颐和园，建成的校园之内却仅有一小片园林，现今的主要教学区——地段所在的白区之内，更是缺少了这样一种休憩的空间。

同时，由于人物设定的原型皆是游戏古剑奇谭的角色，便也借取其主题"问道"，以"固定演员"及"临时演员"的表演之形式，将剧情故事的主要流线编织在园林建筑之中，使游客能够通过表演的观看以及对于园林的亲身体验，在游憩放松的同时，得到一定程度上的"人生体验"。

而在空间的分布上，则选取了游戏中城市、郊野与家园为主要体验内容，分别对应了街道、庭院与居住院落三部分，并依据地形及周边环境，形成了现有功能布局。

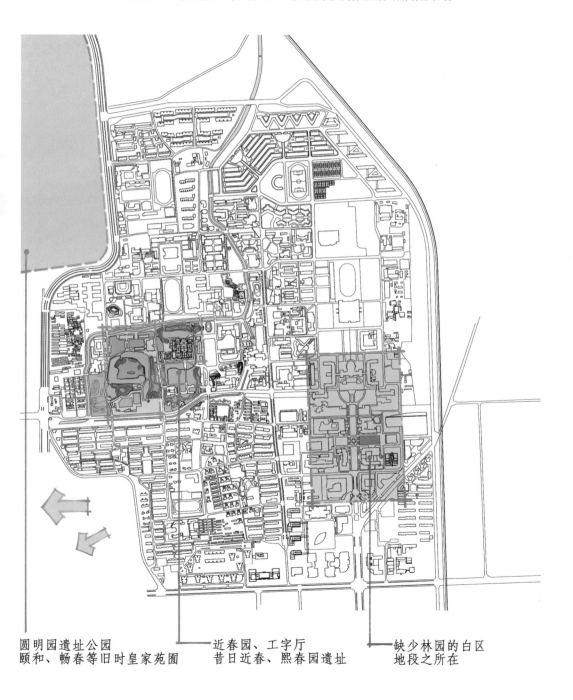

圆明园遗址公园　　　　近春园、工字厅　　　　缺少林园的白区
颐和、畅春等旧时皇家苑囿　昔日近春、熙春园遗址　　地段之所在

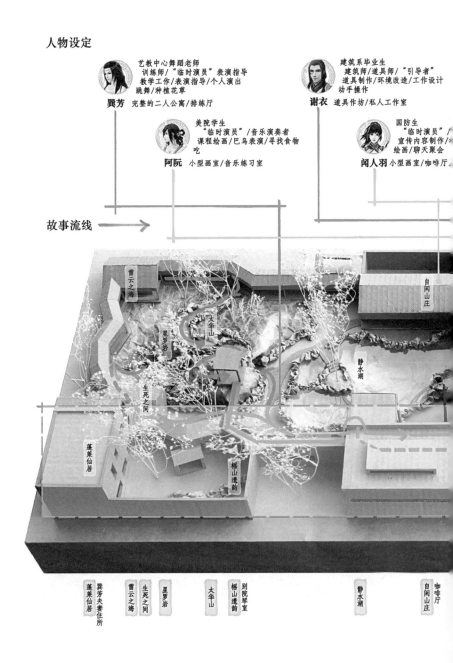

Tsinghua Park is formerly known as the royal garden. Now, although it's still adjacent to Yuanmingyuan Ruins and Summer Palace, there is only a small piece of garden in the built campus left. The present main teaching area, namely the white area of the section, apparently lacks leisure space.

At the same time,, the design integrates the main story line of thevideo game Ancient Swords Miracles into landscape architecture by taking the theme of "ASKTAO", including prototypes of characters, and performance

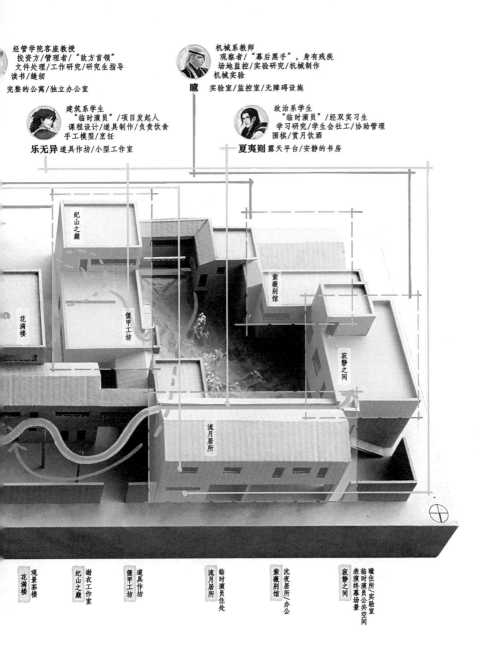

in form of "fixed actor" and "temporary actors", so that visitors can gain a certain degree of "living experience" while relaxing through watching performance as well as the personal experience of the landscape. In the spatial arrangement, it takes the setting of cities, countries and homelands in the game as the main contents of experience, respectively corresponding to the street, courtyard and residential courtyard, and adjusting the existing functional layout according to the topography and the surrounding environment.

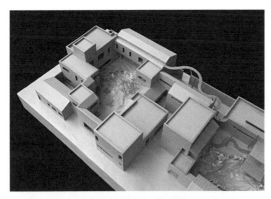

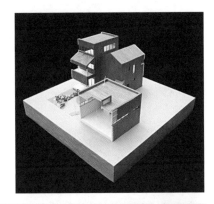

教师点评

刘楚婷的"问道",是将网络游戏"古剑奇谭"中的城市、郊野、家园等虚拟空间转换为分别于现实对应的街道、庭园和居住院落空间,将游戏中的角色原型设定为生活空间中的任务,并将举起故事的主要流线编织在园林之中,使游客能够通过对表演的观看和参与以及在空间中的亲身体验,在居游的同时,得到对"人生问道"的感悟。虚拟与现实的交织与体验,是这一"生活舞台"的重要特色。结合场地条件对各个不同类型空间特别是园林部分的营造都很具特色,设计成果完善,采用纸材料的园林模型和基于模型场景融入的透视图既抽象又表达出生动的氛围。

Teacher's comments

"ASKTAO" of Liu Chuting transforms the cities, suburbs, homes and other virtual spaces in the online game "Ancient Swords Miracles" to the corresponding streets, gardens and residential courtyard space in reality, sets the character roles in the game as the task in the living space, and adds the main story lines to the garden, enabling tourists to obtain the apperception of "life ASKTAO" through watching and participating in the performance and firsthand experience in the space, while living or traveling in the space. Interweaving the virtual experience with reality is an important feature of the "life stage". Creation of different types of spaces, especially the gardens, by combining site condition is very distinctive, and the design results are perfect. The garden model made of paper materials and the integrated perspective drawing based on model scene, though abstract, expresses vivid atmosphere.

本页上图(从左至右):居住院落。假山小亭。茶楼。围护结构研究。
跨页图:恭芳小院。

微社区/MICRO COMMUNITY

项目选址：清华大学建筑学院南侧草坪
项目类型：小型社区综合体
建筑面积：1450m²
用地面积：4000m²

方案设计：逄卓
指导教师：李兴钢
完成时间：2014

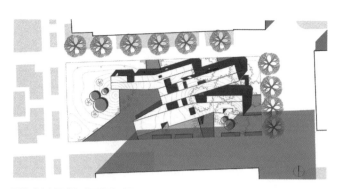

开篇：方案主要透视。**本页上图**：总平面图。**本页下图**：构造剖面和细部。**对页左图**：构成模型。**对页下图**：轴测分析图。

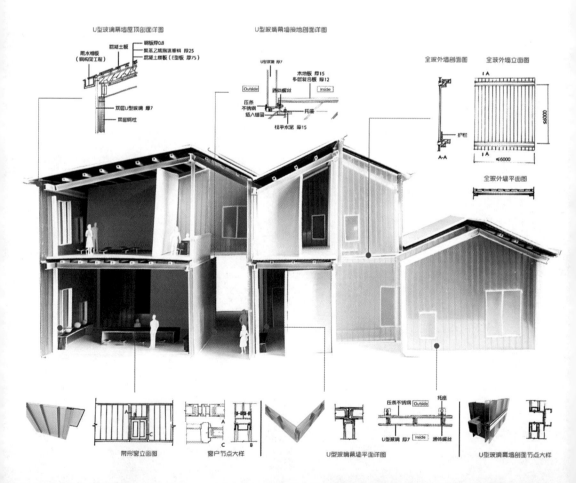

"生活的舞台"这一题目提供了一个有趣的出发点——生活本身。在清华建院南侧的草坪上,我引入了八个熟悉而平凡的角色:退休教授、教师家属、咖啡厅职员等。"微社区"将八个人物(也可是说是八类人物)和学生集合于一个复合的空间内。

设计在某种程度上是对于现有"大院式"小区的反思。现有社区功能单一化、与城市关系封闭化,人们之间的关系日渐疏远。"微社区"没有生硬的边界,与周围的环境有良好互动,场地原本的植物和雕塑被仔细标记并容纳于设计之中。空间的设计既保证使用者私密性的需求,又鼓励人与人之间的交流。钢框架的结构体系保证建筑的轻盈,U型玻璃所带来的半透明效果为公共空间带来明亮开敞的环境。

"微社区"是一个实验性的存在,它以高度的集合性去回应人们复杂多元的生活。生活与建筑并不是简单的一一对应,设计师也不能够从生活演绎推理出一个最完美的空间。在我看来,"微社区"尝试解决一些现实的问题,同时为人们提供一个平台,让生活去定义社区的未来发展,去丰满此处的场所精神。

2014.3.13
街道"家的原型"

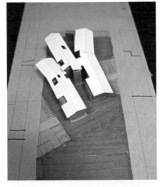

2014.3.18
街道/院落
统一的屋顶形式

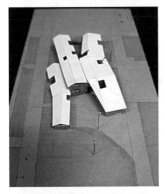

2014.3.20
公共/私密
户外空间的形状

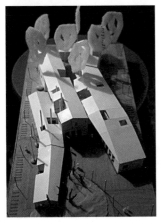

2014.3.22
结合景观/视线
中期评图

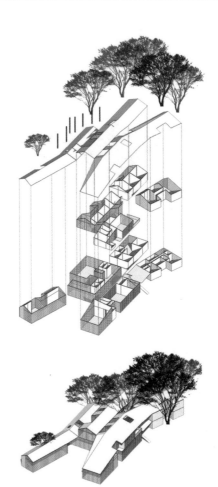

The theme "stage of life" provides an interesting starting point -- life itself. On the lawn of the south side of the School of Architecture, Tsinghua University, I introduce eight familiar and ordinary roles: retired professor, teacher's family, coffee shop staff, etc. "MICRO COMMUNITY" collects the eight characters (also called the eight types of characters) and students in a complex space. To some extent, the design is a reflection of the existing "courtyard style" community. The existing community function is unitary, and the relationship with the city is closed, the relationship between people is gradually alienated. "MICRO COMMUNITY", without rigid boundaries, has good interaction with the surrounding environment, and the original plants and sculptures of the

本页及对页图: 构造模型局部透视拼图。

site are carefully labeled and contained in the design. The design of the space not only ensures the privacy of users, but also encourages exchanges between people. The steel frame structure system ensures the building is light. The translucent effect of the U-shaped glass brings bright and openness to the public space. "MICRO COMMUNITY" is an experiment, taking a high degree of collective responce to people's complex and diverse life. Life and buildings are not a simple one-to-one correspondence, and designers are not able to deduce a perfect space from life. In my opinion, the "MICRO COMMUNITY" tries to solve some practical problems, while providing a platform for people to define the future development of the community and enrich the spirit of this place.

教师点评

逄卓的"微社区",设定了在清华校园内生活工作的8位典型人物及其相互关系,从而以他们的生活内容为基础,展开对场地中建筑和空间的合理想象。她设计了一组介于小型聚落和复合建筑体之间的坡屋顶建筑和丰富的室内外空间,从而与场地乃至校园产生密切和生动的关联。强调内外空间的不同关系,并促进空间中人与人的多样交流,从而成为一个校园中活生生的"生活舞台"。设计的整个发展过程中都能保持清晰明确的概念和意图,不断深化,并通过典型大样空间的研究,使得建筑的结构、材料及构造也得到较为周详的考虑,并较好地体现在建筑的整体效果之中。

Teacher's comments

"MICROCOMMUNITY" of Pang Zhuo is set based on 8 typical characters living and working in the campus of Tsinghua University and their relations. Based on their life contents, reasonable imagination is spread for building and space in the site. She designs a group of buildings with sloping roofs, abundant indoor and outdoor spaces between small settlements and building complexes, so as to generate a close and vivid correlation with the site and even the campus. She emphasizes different relations of inside and outside spaces that promotes communication of various modes between people in the space, forming a vivid "life stage" in the campus. In the whole development process of design, it maintains clear and definite concept and intention, continuously deepen the design, and consider in detail regarding the building structure, materials and construction through the research on typical details, which shall be properly reflected in the overall effect of the building.

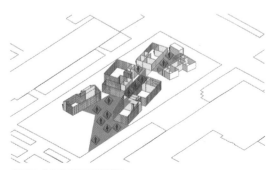

跨页图:构造模型局部透视拼图。
本页下图:私密分级。

梦工厂/DREAM FACTORY

项目选址：清华大学建筑馆南侧草坪
项目类型：社团活动中心

方案设计：蔡宙燊
指导教师：李兴钢
完成时间：2014

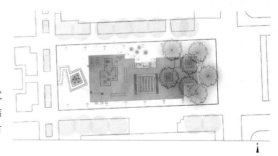

方案主要围绕"梦的社团"的八位人员，根据社团的运行模式以及人物需求布置功能，并由此引发建筑的形态变化，同时结合场地现状，保留场地原有记忆，并留给校园其他人员一个有景观性质的广场。

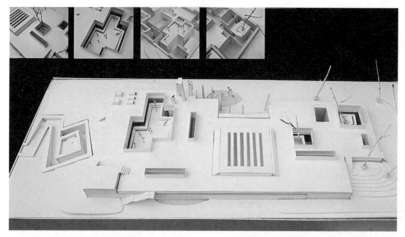

光之睡眠实验室：在实验者入睡前后随着时间的推移，自然光会从不同角度以不同亮度入睡睡眠实验室，是实验者在睡眠时能得到光线的刺激。
风之睡眠实验室：睡眠实验室有高高的捕风塔，能给睡眠实验室提供源源不断的风，使得实验者产生与室内不一样的感觉，对实验者施加风的刺激。
花之睡眠实验室：实验者需穿过一片花丛才能达到睡眠实验室，而在睡眠时也能闻到若隐若现的花香，从而给实验者带来甜美的温馨的梦境体验。
声之睡眠实验室：实验者需穿过消声室才能达到睡眠实验室，睡眠实验室的屋顶为不规则形，能提供良好的听觉效果，使得实验者在美妙的音乐中睡眠。
水之睡眠实验室：实验者需涉水才能达到睡眠实验室，水对足部的刺激以及水的反射投影给实验者带来与众不同的感受，能对实验者的梦境产生影响。

微城市/MICRO CITY

项目选址：清华建筑学院南侧花园
项目类型：校园综合体
建筑面积：850m²
用地面积：4000m²

方案设计：窦森
指导教师：李兴钢
完成时间：2014

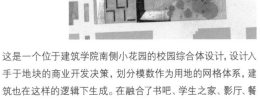

西立面图 1：200　　　北立面图 1：200

这是一个位于建筑学院南侧小花园的校园综合体设计，设计入手于地块的商业开发决策，划分模数作为用地的网格体系，建筑也在这样的逻辑下生成。在融合了书吧、学生之家、影厅、餐厅、工作室等功能后，剖面上也做了错层来满足不同空间高度、不同面积大小的功能使用。此外，为了和基地形成更好的对应关系，在地面层做出通过式走廊以强化人的视线和行为上的联系，其低建筑密度策略同时也换来了更多的建造可能。

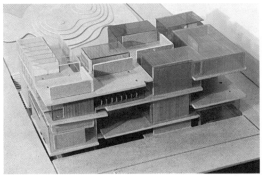

Open Design Studio 2014

变形虫/AMOEBA

项目选址：清华建筑学院南边的绿化用地
项目定位：学生及校友活动
建筑面积：1500m²
用地面积：2100m²

方案设计：金容辉
指导教师：李兴钢
完成时间：2014

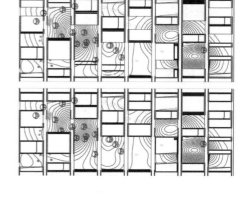

平面图中，可以看到根据活动规模大小而发生变化的空间。通过隔板有限制的活塞运动，使得空间的大小出现变化。图中是大型空间最多的情况与小型空间最多的情况。

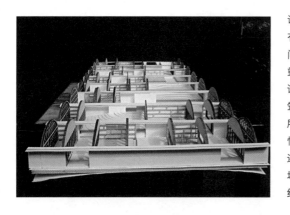

设计者在平时观察中，发现学校有各项大活动开办时，虽然有具有规模的团体开办活动的空间，但是没有小型的、尺度宜人的空间提供给不常用空间的群体。因此，在尽量保留绿化的情况下，希望打造出一个空间大小多变、功能多变的场所。因此，以本次设计的8位同仁为典型的学生形象，做出了八条各具特色的空间。建筑里，功能不再是永恒的因素，随着社会进步速度的加快，我们所需的功能也随之变化。建筑要应对变化，但是不能失去它的本性，因此，需要在永恒的样式上加载多变的功能，样式简单实用，还要体现本地的特征。该方案是清华大学的校友与学生的活动场所，里边加入了曾经在校学习的人的特征亦代表各种不同的人经历过大学生涯。

猫咪咖啡馆/CAT CAFE

项目选址：清华大学建筑学院南草坪
项目类型：工作 / 学习 / 居住 / 游憩
建筑面积：1130m²
绿化率：54%

方案设计：田莱
指导教师：李兴钢
完成时间：2014

基于清华园内有很多流浪猫游荡在宿舍区、图书馆、设计地段等绿地的现状，拟在地段内设计一座猫咪咖啡馆，将这些流浪猫收养起来。同时地段位于主楼与东门之间，人流量大，游客众多，咖啡馆也可以为游客和附近系馆的学生提供休憩、聚会的场所。基于地段现状和功能需求，设计最主要的想法是想做一个人猫共处的综合体。将猫放在与人同等的地位上考虑人的工作与居住、猫的活动与居住空间之间的关系。主要通过层高和连续性的控制创造出只有人的空间、只有猫的空间和人猫共处的空间。另一方面，通过家具和结构系统结合的设计为猫的休息、跳跃提供了空间，也营造出了特殊的空间体验。

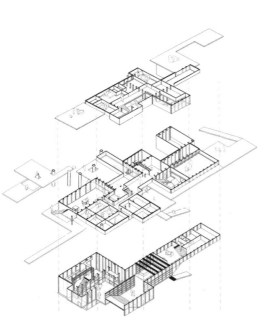

艺圃/GARDEN ART

项目选址：清华大学建筑馆南侧紫薇园
项目类型：艺术家工作室
建筑面积：2320m²
用地面积：4000m²

方案设计：王冉
指导教师：李兴钢
完成时间：2014

总平面图　　0　6　12　　30m

地段毗邻建筑馆与清华美院，具有浓厚的艺术氛围，因此将该地块打造成一个艺术家综合体，服务在校内任教的作家、音乐家、画家与建筑师，满足其居住、工作、游憩、聚会等需要，同时构建不同人群的交流空间，形成"艺圃"。

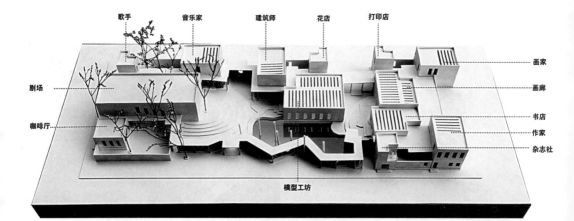

心灵院/MIND HOUSE

项目选址：清华大学建筑学院南边空地
功能定位：心理治疗中心
建筑面积：1500m²
用地面积：4000m²

方案设计：余乐
指导教师：李兴钢
完成时间：2014

根据清华大学学生的心理问题，设想出8个人物，每个人物拥有各自独特的特质和性格。设计主要是为这8个人物提供一个心理治疗场所，取名为心灵院。

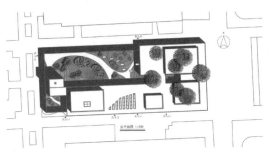

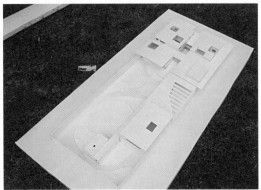

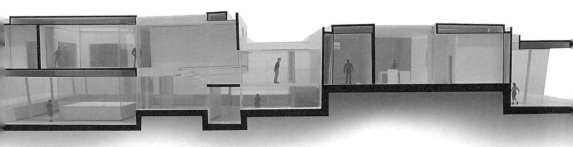

1KM-HIGH

张轲

标准营造事务所 创始人、主持建筑师

城乡聚落

张轲
标准营造事务所
创始人、主持建筑师

教育背景
1988年 – 1993年
清华大学建筑学院 建筑学学士
1993年 – 1996年
清华大学建筑学院 建筑学硕士
1996年 – 1998年
哈佛大学设计学院 建筑学硕士

工作经历
2001年至今
标准营造事务所 创始人、主持建筑师

主要论著
《共生与更新 标准营造"微杂院"》[J]. T+A 时代建筑，2016/07
《西藏林芝南迦巴瓦接待站》[J]. T+A 时代建筑，2009/01
《青城山石头院》[J]. 建筑学报，2008/07
《由外及内 — 中国当代，荷兰鹿特丹》[J]. 时代建筑，2006/05
《阳朔不知名小街上的店面》[J]. 世界建筑，2005/10
《我们希望打破设计的界限--标准营造事务所访谈》[J]. c_d工业设计，2005/04
《Mit der Geschichte planen》[J]. Garten+Landschaft，2004/12
《"标准营造"-关于北京武夷小学礼堂"WESA"的随想》[J]. T+A 时代建筑，2004/02
《关于北京东便门明城墙遗址公园的规划设计》[J]. 世界建筑，2001/8

设计获奖
2013 – 中国博物馆建筑大奖优胜奖
2012 – 芝加哥国际好设计奖（明托盘, Alessi）
2012 – 智族GQ年度设计师
2011 – 意大利维罗纳国际石材建筑奖
2010 – 美国建筑实录国际十大设计先锋
2010 – 车尔尼科夫建筑奖国际十位优秀建筑师特别提名
2010 – 获邀参加威尼斯双年展奥迪城市未来奖展览
2010 – WA中国建筑奖，优胜奖
2008 – 获第一届中国建筑传媒奖，青年建筑师奖
2006 – WA中国建筑奖优胜奖

代表作品
北京东便门明城墙遗址公园（图1）、西藏雅鲁藏布江小码头（图2）、西藏林芝娘欧码头（图3）、西藏林芝南迦巴瓦接待站（图4）、西藏林芝大峡谷艺术馆（图5）、西藏尼洋河游客中心（图6）、西藏尼洋河观景台（图7）、万科回龙观（图8）、北京武夷小学礼堂（图9）、成都青城山石头院（图10、图11）、杭州肖峰艺术馆（图12）、微胡同（图13）、微杂院（图14）、诺华上海研发园区办公楼（图15）

1KM-HIGH城乡聚落

三年级建筑设计(6)设计任务书
指导教师: 张轲
助理教师: 王硕 鲍威

中国的城市与乡村

一项简单的研究显示,1996~2013 年中国至少有1.5亿亩农田被饥渴的城市化所吞噬,占中国耕地总量的8%,这意味着永久失去了养活 1 亿人口的粮食产能。从卫星图上可以看到,越来越多的农田被盲目扩张的城市化所破坏,涉及之处如同癌细胞一样无休止地扩张,触目惊心。在此大背景下,中国城镇化进程仍在继续。全国每年只有部分土地用来开发,这显然不能满足日益增长的开发需求。面对这一难题,城市与乡村向上发展成为必然。

城市与乡村的二元论对于理解中国历史之演进至关重要,两者的互动造成了中国当今社会经济之格局。然而随着新的城市化/城镇化模式的出现,对立地去理解两者间的关系已经不合时宜。我们能否将农村生活的惬意带进城市?我们能否在乡村创造城市的便捷?或者我们能构想出一种取二者之长的新兴发展模式?

竖向聚落

竖向聚落是否可以摆脱现在城乡对立的模式?为应对不同发展模式的需求,是否可以从竖向聚落的角度出发来探讨城市基础设施的问题?竖向聚落是否可以具有城市尺度,从而成为可供居住的大型社区?

教学方法

本课程要求在 600m×600m×1000m的空间范围内构思一座竖向聚落,它可以容纳三个自然村落的人口,每户均有与其原宅基地相称的使用面积。

其功能范围涵盖住宅、幼儿园、小学、商业、办公、餐饮等,即普通聚落里人们日常生活所需的功能空间,也可构想与立体农业的结合。

课程前期以 3~4 人为小组构想竖向聚落的结构系统及形态,共同完成调研报告及模型制作;后期各人均需对选定宅基地进行空间设计。学生们通过此课程,既培养其团队合作能力,也提高各人从大尺度空间里进行单体设计的能力。

在整个课程中,每周均会邀请外聘评委(包括建筑师、规划师及理论家)参与评图。

教学安排

第一周
熟悉现阶段中国农村与城市所面临的挑战,关注中国城市乡村问题研究;对竖向聚落的案例研究;以及对未来城市-乡村一体化的可能性研究,并开始构想一座自己梦想中建在山上的农家宅。

第二周
全班确定3~4个整体竖向聚落体系概念方案,形成小组(3~4人)。

第三周
制作 1:500竖向聚落方案模型,确定各人选作独户宅院设计的地块。

第四周
期中评图,要求有体系分析、结构分析、功能分布、单户住宅方案等。

第五、六周
完善住宅单体设计方案,要求制作完成 1:100 模型。

第七周
独户宅院茶室详细设计及其 1:50 局部墙身节点。

第八周 期末评图

假想未来/3D CITY

项目选址：假想城乡结合部，农田之上
项目类型：垂直聚落规划
用地尺寸：600m×600m
建筑高度：1000m

方案设计：殷玥
指导教师：张轲
完成时间：2014

100M HIGH 空间　　　　2D 街区　　　　　　　　　　　　　　　　　2D Block

Density

2D Block & road

2D city

1KM HIGH 空间　　　　3D 街区　　　　　　　　　　　　　　　　　3D Block

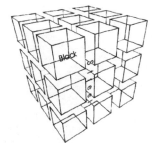

The grow process of A 2D city　　　2D　生长模式　3D　　The grow process of A 3D city

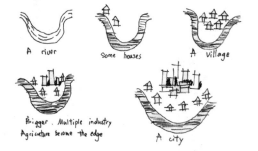
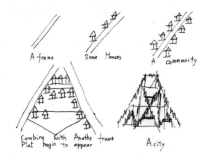

3D城市构思　　　　　　　　　　　　　　　　　　　　　　An idea of A 3D city

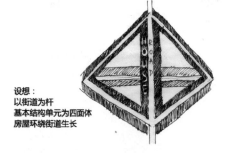
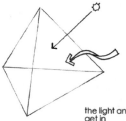

a long time enjoying sunshine shadows will move fast and it feels like cloudy
前杆的阴影在后杆上快速移动
使得后杆同样拥有充足光照
感觉像是多云天气

设想：
以街道为杆
基本结构单元为四面体
房屋环绕街道生长

the light and fresh air get in
引入阳光和新鲜空气

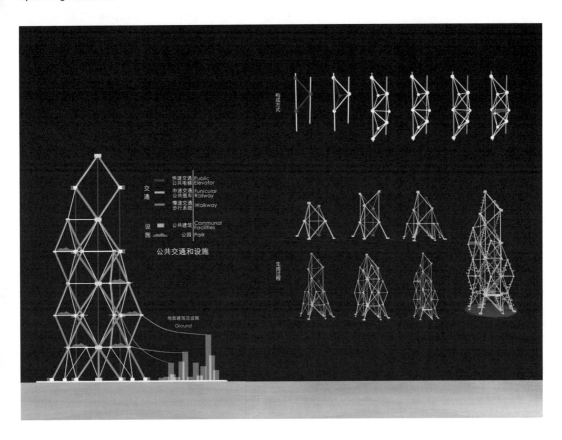

一块600mx600m,高1000m的空间,居住一定的人口,满足他们的需要,并能够提供土地作为农业用地。

那么这个空间会是……

最稳定的三维结构是三棱锥形,在这个以三棱锥为原型生长成的城市结构中,最基础的三棱锥是组成一个立方体的三个三棱锥,新生长的三棱锥是在基础形上的拓扑。

如此的城市结构将会重新定义城市交通和城市街道。每隔150m便有一部作为快速交通的升降梯,在几个主要构造层之间停留;每层之间有"巴士",能够沿着倾斜的结构在杆的两端接驳;较为水平的杆则提供居民悠闲的步行街道。

极致理性的结构,提供丰富的城市街道和公共空间,带来别样的空间体验。

"土地"这一概念将会增加一个维度。在这样一个城市结构中,可建设的"用地"将是一块有特定形状的空间。从主结构上生长的次级结构是建设的建筑结构的基础,"用地"的基准层将和巴士站以及"街道"各层平台结合设置。城市的公共建筑将出现在三棱锥的顶点上,而主要的公园将会设置在升降梯停留的主要构造层,让人们在空中享受植物的拥抱和公园的美景。

开篇:全景效果图。对页及本页图:分析图。

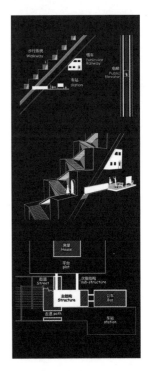

本页上图：分析图。**对页图**：立面图。

If there is a space with a dimension of 600mx600mx1000m (length × width × height) for certain amount of population to live, in order to meet their needs and leave some land for agriculture, this space shall be composed of three-edged tapers, which are the most stable three-dimensional structures.

In this urban structure developed on the basis of three-edged tapers, the most basic part is the 3 three-edged tapers constituting a cube, and the new developed three-edged tapers are topographic part formed on the basis of basic part.

This kind of urban structure would redefine urban transportation and streets. There is an elevator for rapid transit every 150m, and elevators stay among main structural layers; there is a "bus" between two layers, which may be connected to other buses along the inclined structure at both ends of a pole; horizontal poles provide pedestrian streets for residents to enjoy leisure.

The rational structure is composed of abundant urban streets and public spaces, showing a special spatial experience.

The concept - "land" will add a dimension. In this kind of urban structure, constructible "land" would be a space with a specific shape. The secondary structures developed from the main structure are the basis of constructed architectural structures, and the datum layers of "land" would be set with the combination with bus stations and "streets" on each layer platform. Urban public buildings would be constructed on the peaks of three-edged tapers, and main parks would be set in the main structural layers where elevators stay, enabling people to enjoy plants and sceneries.

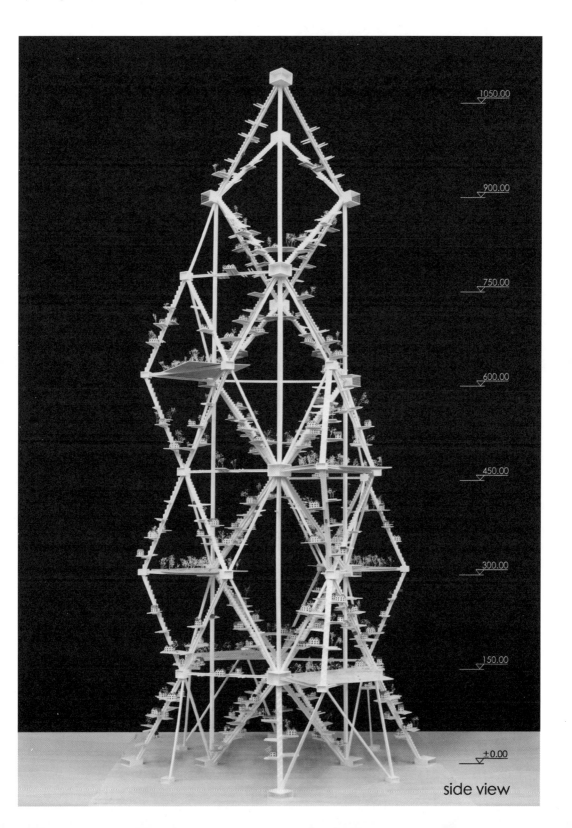
side view

本页上图：手绘概念图。本页下图：俯视效果图。对页：局部效果图。

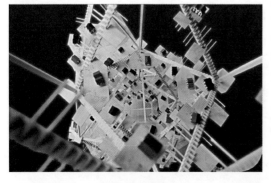

教师点评

殷玥的这一作品给人带来的最大冲击来自于幻景与现实并置时产生的强烈张力：超现实的抽象城市框架与超写实的具象别墅单体，这两个从任何角度都互为极端的事物。

然而，事实却恰好相反。作为幻景的垂直城市，其解决方式却如此真实：空间网架的结构体系、结构交点的交通节点、城市空中公园、垂直以及斜向交通体系等等，设计得恰到好处。一个未来的、巨构的、虚幻的城市落实得如此生动，此为将幻景落回现实的冷静。而作为现实的"别墅"在空中被这样孤零零地展示出来，使其在空间六个方向都和邻居没有任何接触，完全享受其追求的景观与"私密性"。一个个我们熟悉的、"在地"的、人民群众喜闻乐见的房子以如此仙然的方式呈现，此为将现实推向幻境的癫狂。

这一作品看似戏谑与怪诞，背后却是严肃与正统，这比宣言式的提案更加深入人心，恐怕是让大家印象最为深刻的作品。（鲍威执笔）

Teacher's comments

The strong tension produced from the combination between mirage and reality is the maximum impact brought by Yin Yue's design. The surreal abstract urban framework and super-realistic concrete villa are two extremes mutually from any angle.

However, the fact is just the opposite. For the vertical city as a mirage, its solution is so true: The structural system of spatial truss, transport nodes at structural joints, urban hanging parks, vertical and inclined transportation systems, etc. are designed properly. A future, mega-structure, visional city is described so vividly, showing the calmness of implementing the mirage to practice. The "villa" as reality is shown alone in the air without any contact with neighbors in six spatial directions, enabling the residents in it to completely enjoy the landscape and "privacy" they pursued. Those houses which are familiar to us on ground and delighted to the masses are presented in such a way, which is a crazy thought to push reality to migrate.

This design seems joking and weird, but actually it is put forward on the basis of seriousness and tradition. Therefore it is more profound than a declaration proposal, and it may be the most impressive one. (Written by Bao Wei)

1KM-HIGH
垂直聚落/VERTICAL VILLAGE

项目选址：北京城乡结合部
项目类型：城市综合
用地尺寸：700m×500m
建筑高度：845m

方案设计：沈毓颖
指导教师：张轲
完成时间：2014

 方案设计：张孝苇
指导教师：张轲
完成时间：2014

 方案设计：李培铭
指导教师：张轲
完成时间：2014

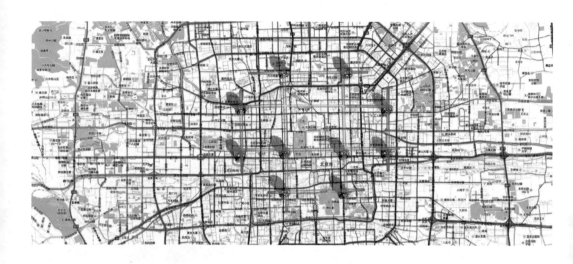

城市问题·高楼困境 随着城市化进程的不断推进，当今的大都市——特别是北京——渐渐在林立的高楼中产生了诸多问题和困境：它们密度极高，在吸纳大量人群的同时缺少足够的公共空间进行"呼吸"；它们尺度巨大，一座座林立的高楼在尺度上并不宜人，给人们以冰冷的疏远感；它们缺乏交流，不同阶层的人群往往没有深层次的交往，逐渐造成社会分层的加剧。 与此同时，诸如北漂、城乡割裂以及人口老龄化加剧等社会问题也在城市化进程中被逐渐放大，这一系列城市问题也被用来指引我们进行垂直村落的设计。

聚落体系·生长机制 垂直村落主要由三个层面的体系构成：位于聚落内部的垂直核心，作为结构核心和交通核心的作用；基于框型社区的大型公共绿地；以及公共绿地之间的流线设计。垂直聚落遵循以下的生长机制：首先设置垂直核心和基础设施"块"，其次设置公共服务"框"，再根据前两者进行居住"空间"的设置与安插，在聚落逐渐生长的过程中外部的"框"逐渐代替内部的"框"，后者则成为新的基础设施"块"发挥作用。

交流 我们在方案聚落层面上着重考虑了大型公共空间的设置，在每一个"框"型社区相互搭接的过程中自然形成诸多平台，我们将其定义为城市尺度上的广场、公园等游憩空间，从而试图解决当前高楼缺乏公共空间的难题；同时利用城市尺度上的大型斜向连接，提供社区到社区的快速通道，完成城市和社区尺度上的交流与交通联系。

尺度与生活 方案在聚落的大中小三个尺度上均有相应设计，大尺度上包括基础设施、公共服务、居住生活区，中尺度上包括空间与空间之间的联系、平台绿地和广场等，小尺度上则着重营造北漂群体社区、老年社区以及田园社区等三种生活意向。设计的最后我们在各个社区内选取了$9m^2$的空间进行了有趣的设计，包括针对北漂群体的$9m^2$屋顶舞台、针对田园社区的$9m^2$茶室，以及针对老年社区的$9m^2$书屋。

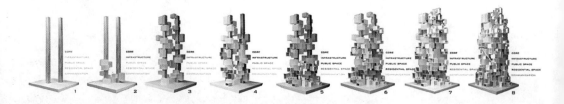

-HIGH DENSITY-

-MASSIVE-   -LACK OF COMMUNICATION-

THE BEIJING FLOATER DILEMMA
Beijing drifters are people who live and work in Beijing, while without Beijing registered residence and change their houses from time to time. They are contributors to Beijing without the oppotunity to enjoy the good living conditions they deserve.

THE SAGREGATION OF CITY AND FARMLAND
With rapid process of urbanization, the segregation of city and farmland is increasingly serious. High density lead to the loss of biological system in city, at the same time, the shedding population in country cause more and more discarded farmland.

THE PROBLEM OF AGING POPULATION
The problem of aging population has become a big challenge for China, the aged people especially the empty-nesters call for a lot of attention and care from the whole society.

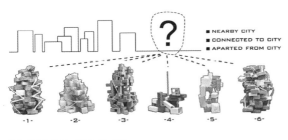

开篇：模型细节图。**本页上图**：设计问题分析。**本页下图**：生成逻辑分析。**对页下图**：生成分析。

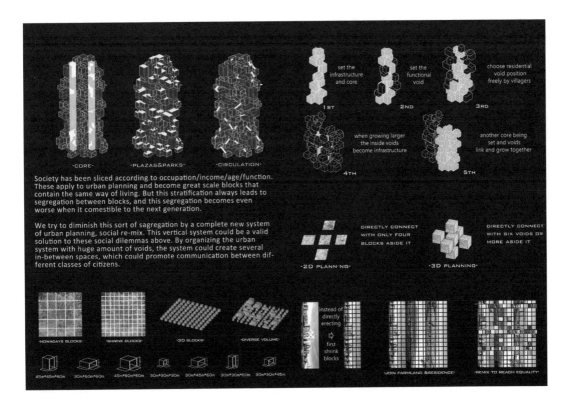

[9 ㎡ STAGE ON THE ROOFTOP]

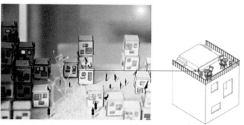

The stage is designed to be on the rooftop of a small house for a typical beijing floating musician, who is struggling for his living by playing music.

The 9㎡ stage can serve as a place for musicians to show their musical talent to the whole community, and as a place for musician to drink and entertain with his close friends.

[9 ㎡ TEA HOUSE]

The tea house locates on the edge of the layered farmlands. It provides an alter path to the upper layer, and a brilliant view for the green plants outside. Sitting on the mat, forone farmland wringgles as a river.

[9 ㎡ BOOK HOUSE]

The 9㎡ is a book house designed for an aged man who is a bibliophile. He can share his books with other young people passing by the book house by putting the books in the showcase.

Moreover, the book house also provide place for children in the neighborhood to read books, study and get taken care of by the aged man. The aged man will also get the oppotunity to communicate with people of dif-

Urban problems, the plight of high-rise buildings:
With advancement of the urbanization process, the modern metropolis, especially Beijing, gradually accumulate many problems with the high-rise buildings: they are of very high density, and lack of adequate public space to "breathe" while absorbing a large number of people; they are huge in scale, and a forest of over scaled tall buildings do not make people comfortable, resulting in a sense of alienation; they lack of communication, that people of different levels often do not allow for frequent communication, which gradually caused the intensification of social stratification. At the same time, drifters in Beijing, urban and rural split, aging population and other social problems are also gradually expanding along the urbanization process. This series of urban problems is also used to guide us in the design of vertical village.

Settlement system, growth mechanism:
The vertical village is mainly composed three levels of system: vertical core located inside of the settlement used as structure core and traffic core; large public green space based on box type community; and streamline design between the public green spaces. Vertical settlement follows the following growth mechanism: first set the vertical core and infrastructure "block", then set public service "box", and place the living "space" according to both items above; the external "box" gradually replaces the internal "box" in the gradual growth of the settlement, and the latter becomes new infrastructure "block" to play it's role.

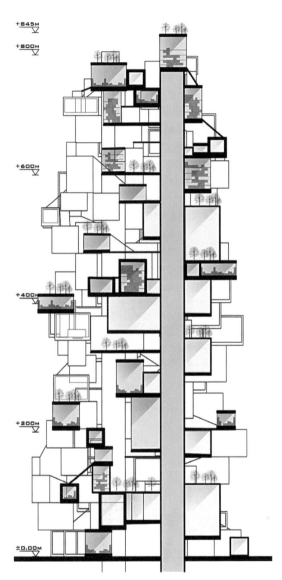

Exchange:

We emphatically considered the setting of the large public space at the settlement level of the scheme. In the process of connecting each "box" type community, many platforms are formed naturally, and we define it as the squares, parks and other recreation spaces in the scale of city, so as to attempt to solve the problem that the lack of public spaces for current buildings; at the same time, using the large oblique connection in the city scale, the fast passage from community to community is provided, thus completing the communication and transport links in urban and community scales.

Scale and life:

The scheme has corresponding designs in the large, medium and small scales of the settlement. The large scale includes infrastructure, public services and residential living areas; medium scale includes the connection between spaces, plat form green space and square; and small scale focuses on creating the communities for drifters in Beijing as well as communities for the elderly and pastoral communities. At the end of the design, we selected 9m2 space from different communities to implement interesting design, including 9m2 roof stage for drifters in Beijing, 9m2 teahouse for the pastoral community, as well as 9m2 book house for community for the elderly.

本页: 原始概念。对页: 主要方案图。

[LARGE SCALE]

STACKING VOID
- INFRASTRUCTURE
- PUBLIC SPACE
- RESIDENTIAL AREA

[MIDDLE SCALE]

SMALL CLUSTERS OF BOXES
- COMMUNICATION BETWEEN BOXES
- PLATFORMS AND ROOFTOPS
- GREENLAND AND PUBLIC SQUARES

[SMALL SCALE]

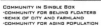

COMMUNITY IN SINGLE BOX
- COMMUNITY FOR BEIJING FLOATERS
- REMIX OF CITY AND FARMLAND
- COMMUNITY FOR AGING POPULATION

教师点评

"落盒子"作为一种形式原型并不罕见,但是当其被应用在"1km"这一城市尺度上时,面临的问题一定是没有任何参考先例的。如何把握盒子单元的尺度,如何安排好功能关系,如何解决结构、交通的问题,都是这个方案成败的关键。虽然此组的出发点是解决社会问题,但当最终落实到建筑形态时,上述问题必须予以解决。本组同学从一开始概念便确立,但面临的挑战并不因此减弱,反而难度系数更大。三位同学经历了很多轮的尝试,有了一定的突破,比如能够将平面城市街块尺度与立体盒子单元的尺度进行联系,对社区生活场景的营造等。但也有很多基本的问题没能够予以充分解决,比如基本的水平向交通等。

我想方案的成功与否或许并不是这次尝试的最终目的。通过这些努力与尝试,能够直面这些建筑面临的基本问题才是最重要的。相信经过这次训练,同学们会对今后实践有相应的准备,也会明白,只有不断地去努力尝试,才能在设计上有更大的突破与超越。(鲍威执笔)

本页左图:模型不同尺度展示。**本页下图**:模型局部展示。**对页**:南立面图。

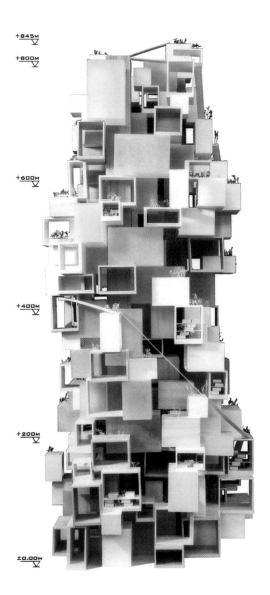

Teacher's comments

It is a common practice to take "stacked box" as a formal prototype. However, when it is applied to "1km", an urban scale, there must be no reference precedent for the problems faced. How to control the dimension of box unit, how to arrange functional relationship and how to solve structural and transportation problems are the key factors deciding the success or failure of this scheme. Although the starting point of this group is to solve social problems, the above problems must be solved in architectural form.

From the beginning, this group of students established the concept, but the challenge they faced did not abate. On the contrary, the degree of difficulty was higher. The three students attempted for several times, and made a breakthrough. For example, they are able to connect plane urban street block dimension and the dimension of vertical box unit, create community living scenes, etc. However, there are some basic problems to be fully solved, such as the basic horizontal transportation, etc.

In my opinion, the success of scheme may not be the ultimate purpose of this attempt. Through those efforts and attempts, it is important for us to directly face those basic problems for those buildings. I believe through this training, students will have corresponding preparation for future practice and understand a principle – Great breakthrough and surpassing can be obtained only through continuous efforts and attempts. (Written by Bao Wei)

折叠的风景/FOLDED LANDSCAPE

项目选址：虚拟地段城郊
项目类型：城市聚落
用地尺寸：600m×600m
建筑高度：1000m

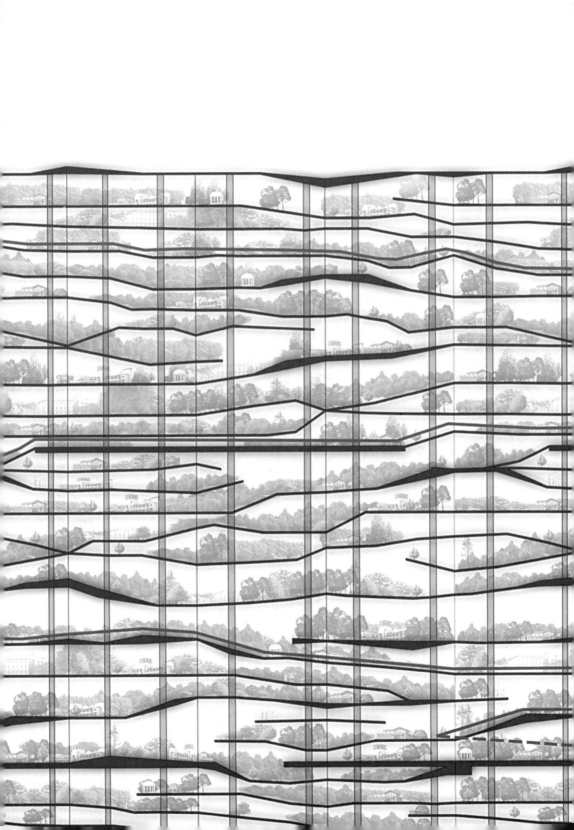

方案设计：马宁
指导教师：张轲
完成时间：2014

方案设计：蔡忱
指导教师：张轲
完成时间：2014

开篇：步行螺旋上升的路线。**本页图**：生成过程。**对页**：效果图。

Chinese Habitats　　**Extracted Landscape**　　**Continuous Shape**　　**Funcion Based on Landscape**

我们的设计主题是Folded Landscape，希望在1km高度的聚落内营造丰富的地景空间形态，形成在城市和乡村中不断"穿越"的生活体验，打破现在的城乡二元对立结构，重新定义城市和乡村的关系。1km聚落位于城乡结合部，首先以600mx600m为一个平面单元，分别选取中国城市和乡村中有特别记忆的空间片段，其中包括城市的居住区、大学城、CBD等等，以及具有田园特征的，如北方的四合院、江南的园林等等，还有自然中的山丘、湖泊、森林等要素。然后将600mx600m 的用地在1km高的空间内复制为25层，并在层与层之间通过剪切、折叠、连接等多种方式加强功能之间的互动和联系。1km聚落的主体结构为五根巨型的核心筒，同时承担快速的垂直交通功能和市政基础设施的铺设。

Our design theme is Folded Landscape. It is our hope to create rich landscape space within 1km height of settlement, form continuous "interweaving" experience of life in urban and rural areas, break current urban-rural dualistic structure and redefine the relationship between urban and rural areas. 1km settlement, located in the rural-urban continuum, first takes 600M*600M as a planar unit, which is full of the space segments with special memories in China's urban and rural areas, including the city's residential areas, University City and CBD. Also, it includes the places with rural characteristics such as quadrangles in the north and gardens in the regions south of the Yangtze River, as well as of the hills, lakes, forests and other elements in the nature. Then, the 600M*600M land is replicated by 25 layers within 1KM high space. The interaction and relationship between the functions are enhanced by shearing, folding, connection and other treatments for the layers. The main structure of 1KM settlement is five giant core tubes, which undertakes the rapid vertical transport function and the laying of municipal infrastructure.

对页：立面图。**对页下图**：内部效果图。**本页上图**：不同种类的地形图。
本页下图：内部效果图。

本页及对页上图: 内部意象图。**跨页图**: 内部效果图。

教师点评

这组设计让我很感兴趣的是他们对"田园都市"的处理模式，即通过一种"Sampling"（采样）的方式重新建立了一种"乡愁"的叙事可能。其实"创世纪"中的诺亚方舟，为了能在大洪水之后重建文明，搭载了各种飞禽走兽，也是用"采样"的办法；而城市的疯狂扩张，对于原本承载着自然的土地，又何尝不是一场人为的洪水呢？

设计从另一个角度证明，未来的聚落不一定是要通过人为"设计"出来的，也可以是设置一个框架，即由5个核心筒串起来的25片地景，而它的组织方式又完全是以"城市"基础设施为依托的。然后在这一巨型基础设施上由人们按照某种标准来选择典型的城乡"片段"进行收藏和移植，这样人们集体意识中的田园生活就自然地在"Folded Landscape"上展开了。

最终，当人造的片段在剖面连接上成为连续不断的都市-田园景观时，似乎暗示了另一种意味的"千里江山图"。（王硕执笔）

Teacher's comments

The interesting part of this set of designs to me is their handling mode to "garden city", that is, a narrative probability of "homesickness" is reconstructed through "sampling" method. In order to reconstruct civilization after flood, the Noah's Ark in "Genesis" carried various birds and beasts, which meant the same "sampling" method; the crazy expansion of cities to the land originally bore with nature is a "artificial flood"?

From another angle, the design proves that it is unnecessary to artificially "design" future settlements. A framework may be set, which means, 5 core tubes link 25 landscapes, and the framework's organization form is completely based on an "urban" infrastructure. People select typical urban & rural "fragments" to be collected and transplanted according to certain standard on the mega infrastructure. In this way, the pastoral life in people's collective consciousness is carried out naturally on "Folded landscape".

At last, the artificial fragments connected on the profile become the continuous urban – pastoral landscape, which seemly suggests a "boundless view" with another meaning. (Written by Wang Shuo)

在生活中修行/
PRACTICE IN LIFE

项目选址：山野
项目类型：冥想
用地尺寸：200m×300m
建筑高度：1000m

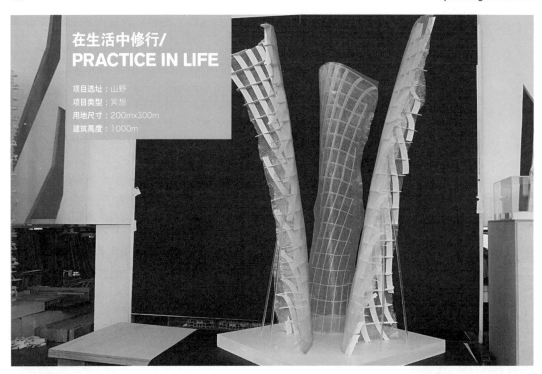

方案设计：张乃冰
指导教师：张轲
完成时间：2014

这次跟随张轲、王硕和鲍威三位老师的设计真的是收获特别大，这个1KM HIGH的垂直聚落设计本身听起来是件挺天马行空的事，确实我们在设计过程中也没有太多的束缚，大家都玩得比较自由随心，也激发了我们每个人内心深处喜欢建筑的天性。几位老师在百忙之中仍然坚持每周给我们非常多有用的指导建议，而且各种有趣的案例都为我们打开了设计的新世界的大门，特别是最后自己要动手做一个大比例的模型，老师们也都提供了非常多实质性的帮助，让我们不但收获了浪漫的建筑理想，也收获了实际的动手能力，感谢老师。

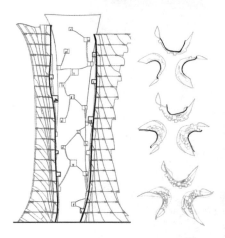

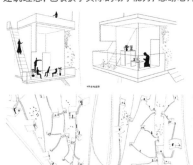

Open Design Studio 2014

垂直村落/
1KM-HIGH VERTICAL HABITAT

项目选址：城市之外，山村之间
项目类型：大型社区综合体
用地尺寸：600m×600m
建筑高度：1000m

方案设计：许东磊
指导教师：张轲
完成时间：2014

方案设计：徐滢
指导教师：张轲
完成时间：2014

随着近年来中国快速的城镇化进程，越来越多的人们离开乡村进入城市。人们在享受便利的同时却必须面临城市的喧闹与拥挤。于是对乡村生活与自然风景的向往成为了都市人的一种普遍情怀。我们类比平原上"一路串多村"的模式，试图通过这个1km高的垂直村落，既满足城市高密度的空间需求，同时又能为人们提供亲近自然的田园生活。

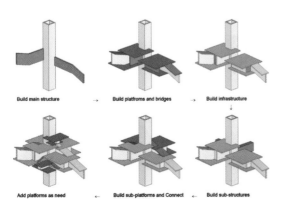

飘 / LIGHTWEIGHT NETWORK

项目选址：城市中
项目类型：生活
用地尺寸：500m×500m
建筑高度：1100m

方案设计：刘东元
指导教师：张轲
完成时间：2014

方案设计：杨良崧
指导教师：张轲
完成时间：2014

设计初期概念阶段，构想了一种相对孤独的、出世的生活态度，来体现对"城市——乡村"这一课题的思考，所谓"大隐隐于市"。设计深化过程中，为贯彻这种能更多时间留给面对自己的生活方式，有意让结构以及对应的交通方式更加写意，营造每个人都能够仰望天空的冥想空间。

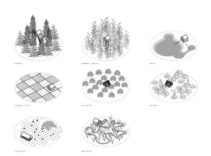

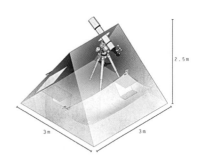

极细聚落 / A THIN HABITAT

项目选址：乡下
项目类型：新农村建设
用地尺寸：200m×200m
建筑高度：1000m

方案设计：陈晓东
指导教师：张轲
完成时间：2014

该设计源于现代城市和传统聚落的肌理对比，探讨邻里关系在塑造聚落生态、情感的积极作用。在用地短缺的极端条件下，尝试以宅基地为概念，试图在垂直维度重构未来的邻里关系，以维系传统聚落的活力。

1
A typical rural habitat
In northeast China

2
Divide the habitat
by the homestead
boundary

3
Rearrange
homestead
in vertical.

建筑

邵韦平

北京市建筑设计研究院有限公司
执行总建筑师

邵韦平
北京市建筑设计研究院有限公司
执行总建筑师

教育背景
1980年 – 1984年
同济大学 建筑学 本科
1992年 – 1996年
同济大学 建筑设计与理论 硕士研究生

工作经历
1984年–1995年
北京市建筑设计研究院有限公司
助理建筑师、建筑师
1995年–2001年
北京市建筑设计研究院有限公司
高级建筑师、国家一级注册建筑师
2002年至今
北京市建筑设计研究院有限公司
执行总建筑师兼院方案创作工作室主任
2010年至今
清华大学建筑学院 硕士导师
2002年至今
北京建筑大学 硕士导师
2013年至今
中央美术学院建筑学院 硕士导师
2010年至今
中国建筑学会 常务理事
2009年至今
中国建筑学会建筑师分会 理事长

主要论著
《BIAD 建筑设计深度图示》，中国建筑工业出版社出版，
2010年
《BIAD 建筑专业技术措施》，中国建筑工业出版社出版，
2005年
《凤凰中心建筑创作与建筑当代性思考》，深圳 中国建筑

学会2014 年年会"多学科融合的建筑设计创新"学术论坛，主题报告人，2014/11
《基于整体建构与数字技术的现代性表达》，《建筑学报》，2014/5
《面向未来的枢纽机场航站楼—北京首都机场T3 航站楼》，《世界建筑》，2008/8

设计获奖
2016 – 北京凤凰中心获得"第十三届中国土木工程詹天佑奖"
2015 – 北京凤凰中心获得"亚洲建协建筑奖（荣誉提名）"
2015 – 北京凤凰中心获得"中国建筑学会建筑创作奖，金奖"
2015 – 北京奥林匹克公园中心区下沉花园获得"全国优秀工程勘察设计银质奖"
2014 – 北京凤凰中心获得"中国勘察设计协会"创新杯"建筑信息模型（BIM）设计大赛最佳BIM 建筑设计一等奖"
2014 – 北京凤凰中心获得"WA 中国建筑奖，技术进步优胜奖"
2013 – 中国尊(CBD-Z15)获得中国勘察设计协会"创新杯"建筑信息模型（BIM）设计大赛最佳BIM建筑设计一等奖"
2011 – 北京凤凰中心入选罗马中意文化交流展核心参展项目获Designboom.com 世界十大文化建筑
2012 – 北京奥林匹克公园中心区下沉花园获得"亚洲建筑师学会金奖"

代表作品
北京凤凰中心（图1、图2）、北京奥林匹克公园中心区下沉花园（图3）、北京首都国际机场T3航站楼(中外合作项目)（图4）、华侨城一期（图5）、中国尊(CBD-Z15)（图6）、CBD核心区公共空间（图7）、奥南商务中心区公共空间（图8）

建筑师之家

三年级建筑设计（6）设计任务书
指导教师：邵韦平
助理教师：刘宇光 李淦

课程目标
1. 研究基地与周边场所的环境关系，寻找创意的方向。
2. 研究潜在使用的行为特征，构建功能关系图解。对当代社会文化和技术特征提出自己的创新见解。
3. 根据研究结论，完成建筑师之家的方案设计，为建筑师和相关领域的设计师、艺术家等"泛建筑师"行业执业者提供工作和交流的场所。

项目基地环境
1. 基地位于北京妫河创意产业园，此园区地处北京延庆县妫河北岸延庆镇西屯村西南地块，占地面积21公顷，总建筑面积14.5万m^2，是一个以建筑创意产业为主，形成设计研究、教育培训、展示交流的创意平台。
2. 园区城市设计意在通过建筑创意园的规划建设提高延庆的环境吸引力，将基地打造成一个湿地、林地、坡地、台地为一体的郊野公园，采用即插即用的个性化模块单元，灵活变化的"簇群式"功能组合，使未来创意创意产业园成为一个开放的创意服务平台，兼容各种个性化团队的需求。力争将项目建成为具有环境吸引力、功能吸引力、空间吸引力、生态吸引力 和文化吸引力的场所。
3. 本项目位于园区01-06-22 地块，用地面积2957 m^2，见附图。
4. 建设场地地形、地貌：建筑创意区现状地形比较平坦，东北角有低洼地块，南侧有堆土，比周边场地高4 m左右，整个场地标高中间较高，东西较低，北高南低，最大标高为485.53m，最低标高480.60位于东北角低洼处。本项目位于地块的西南，该地块现状地势平坦，高差不大，平均标高为484.80，邻接场地标高较低，东西两侧标高为484.50m，南侧紧邻意大利花园用地，其设计标高较低为484.45m。
5. 周边交通条件：建设用地西北侧为创意区红线9.0m宽内部道路，路面宽度为5.0m，人行道宽度两侧各2.0 m，便于自驾车和消防车通行，人行交通便利。
6. 相关信息参见1:3000 园区总平面图(A3)、1:3000 园区地块总平面图(A3)、1:500 基 地总平面图(A3)。

建筑规模和功能要求

1. 建筑面积3600 m²,包含办公、展示、交流、餐饮和住宿功能,需满足以下使用要求:

A. 2 个15 人设计团队的工作、会议和讨论空间。

B. 不少于100m²的展示空间。

C. 供100使用的演讲、演示场所。

D. 供50人就餐的餐厅和厨房。

E. 15~20间标准客房。

F. 以上空间可考虑共用、借用等灵活使用的方式;可相应配备酒吧、观景平台等功能。

G. 满足以上功能的服务用房和设备机房。

2. 园区已设置集中停车场,本地块仅设5个临时停车位。

规划要求

1. 规划总用地面积: 2956.9m²

2. 建筑控制线: 退红线1m

3. 建筑限高: 9/12m

4. 容积率: 0.8

5. 绿地率: 30%

6. 地下室外墙轮廓不得超出首层外墙轮廓

设计成果

1. 环境分析报告

2. 行为、功能关系分析报告

3. 建筑师之家的方案设计

PLASMA建筑师工作营/ARCHITECT WORK CAMP

项目选址：北京坝河创意产业园
项目类型：办公、居住
建筑面积：3267m²
用地面积：2956m²

方案设计：马逸东
指导教师：邵韦平
完成时间：2014

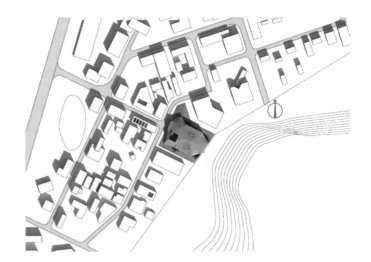

2014.3.13
街道"家的原型"

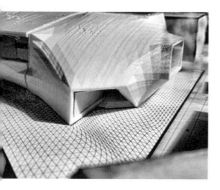

策划时提出了建筑师工作营这种行为模式,两个设计小组像封闭集训一样工作与生活在这座建筑里进行方案突击,这种特殊的行为模式是对当下快速设计行为的极端体现和反思。崇尚工作而摒弃休息,内部密切交流而与外部隔绝,两个小组间的竞争关系,这种极端对立的情景成为建筑设计的概念。

研究这种行为模式并结合场地分析设计剖面与平面,布置功能块。例如将开放无柱的工作空间放于上层而狭小拥挤的居住功能位于下层,这种特殊的排布方式适应了这种极端设计情景。

造型需要体现这种极端的情景,让人直接感受到激烈和挣扎的情感。功能到造型的转译受了Cocoon Art的启发,使得功能块看上去如同在一个紧绷的网中挣扎。模型中使用力学找形模拟这样的表皮。

结构选择了钢梁排架满足大跨的需求,还设计了墙身细部大样以及室内布置实现开始的设想。

开篇:主要方案图。**对页上图**:总平面图。**跨页图**:主要方案图。**本页上图**:入口。**本页下图(从上至下)**:一层平面图。二层平面图。

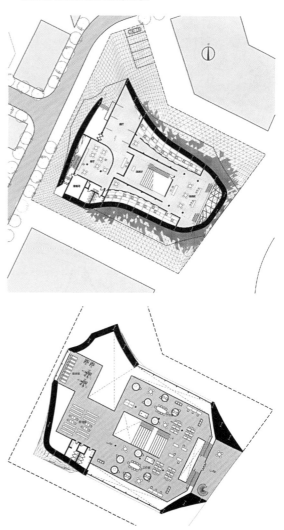

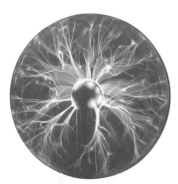

In planning, program of architecture workshop is proposed, where two design groups work and live in the building for to brainstorm schemes like an enclosed training camp. Such program is the extreme expression and reflection on the current design methodology. Such strong advocation for work that abandons resting, internal intense communication vs isolation to the outside, as well as the competition between the two groups, contributes to the extreme disturbing scenes that is embedded in the concept of architectural design. Based on study of the behavior pattern of such program, we analyzed the design's section and plane in relation with the site and arranged program massing blocks accordingly. For example, the open column-free work space is set at the top and the narrow crowded living functions are set at the bottom. This special arrangement adapts to the extreme design scenario. The forms need to reflect this extreme scenario, enabling a person to directly feel the intensified and struggling emotions. The translation from function to the form is inspired by the Cocoo Art, making the functional massing block seem to be struggling in a tightened net. In the model, the mechanics form finding is to simulate the surface. The structure is the reinforced concrete trestle, which meets the needs of large span; the wall details and interior layout are also designed to realize the initial concept.

本页上图：概念图。**本页下图**：场地分析图。**对页上图（顺时针）**：功能概念。功能类型比较。功能定位研究。**对页下图**：功能重新分组。

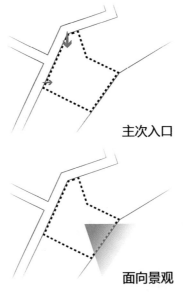

主次入口

面向景观

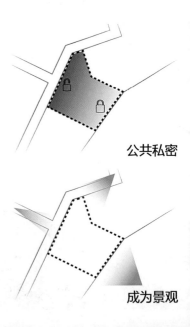

公共私密

成为景观

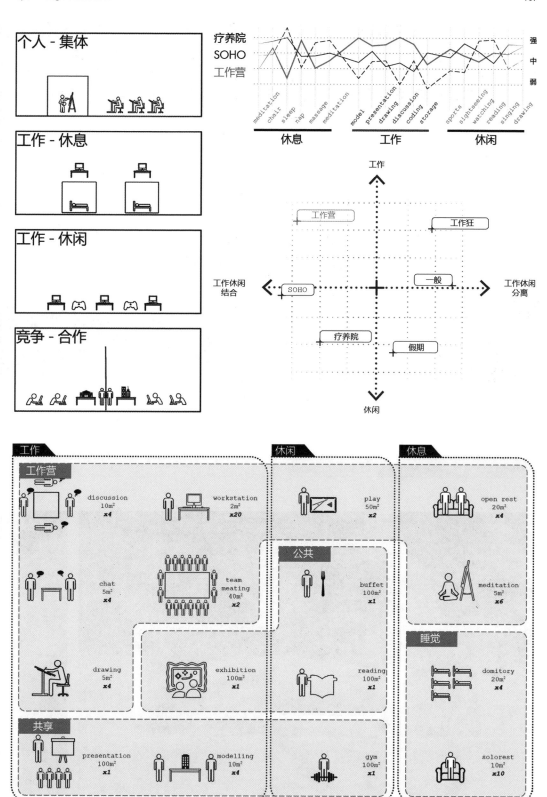

教师点评

马逸东同学是学校推荐的课程设计组学生召集人,在设计课中一直表现出良好的专业状态。马逸东对任务书进行了深入的解析,通过对设计师工作与生活规律的分析与研究,用等离子体紧张、冲突、矛盾的概念来比喻设计师的创作状态,用蚕茧的概念来表现建筑,因而方案创作走向非线性方向。在设计中他没有仅仅停留在形式塑造上,对建筑的设计逻辑、结构体系、维护材料都有完整的应对。作为大三年级的学生能够根据自己的思考,在老师的指导下完成如此复杂的课程设计实在难能可贵。当然由于受课程时间和经验的局限,作业中有些内容还不够完善是可以接受的,希望马逸东同学保持良好的学习状态,取得更大的专业进步。

Teacher's comments

Ma Yidong is the person coordinating students for design team of the course as recommended by the school. With a good professional attitude in the design course, Ma Yidong throughout analyzes the assignment brief, compares creative attitude of the designer to the concepts of tension, conflict and contradiction of plasma through analysis and research on working and living laws of the designer, and presents the building with the concept of silkworm cocoon, and made the creative move towards non-linear direction. In the design, he not only stays at form finding, but completely copes with design logic, structure system and maintenance materials of building. As a junior student, it's really estimable to complete such complicated course design based on his own thinking under the guidance of teachers. Certainly, limited by course time and experience, it's acceptable that there are imperfect contents in the work. I hope that Ma Yidong can keep the positive learning attitude, and gain further professional progress.

跨页上图:内部效果图。**跨页下图**
(**从左至右**):从河岸看。街景之一。

对望·芦苇荡/LOOK, THE REEDS

项目选址：北京坊河创意产业园
项目类型：办公、居住
建筑面积：3700m²
用地面积：2956 m²

方案设计：孙冉
指导教师：邵韦平
完成时间：2014

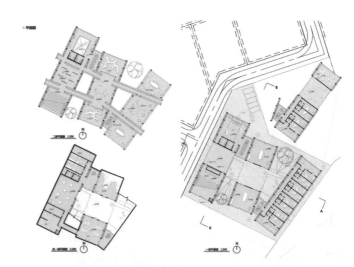

开篇：主要方案图。**本页上图**：平面图。**本页下图**：主要方案图。**对页**：立面图与剖面图。

建筑师之家的地段处于北京的西北妫河创意园区。功能定位于举行建筑师夏令营，集合了居住、工作、交流等功能为一体，将展览交流等人流引向容易到达的区域，给建筑师创造更大的活跃度和自由空间。地块从西北到东南的开放性逐渐降低，私密性逐渐提高。东南侧良好的景观朝向和日照使得东南侧的居住休闲舒适程度较高。因此将地块划分为三条，功能对应交流、工作和居住。三块功能之间需要很好的联系。横向的通道吸引着人朝向交流空间和观景平台，在工作和生活空间之中营造出一片建筑师与外界的交流空间。用芦苇杆来作为主要建筑的表皮。这样的建筑空间之间既可以对望，又不失独立性，同时就地取材很利于节约成本和资源，促进材料的合理利用，保护生态环境。

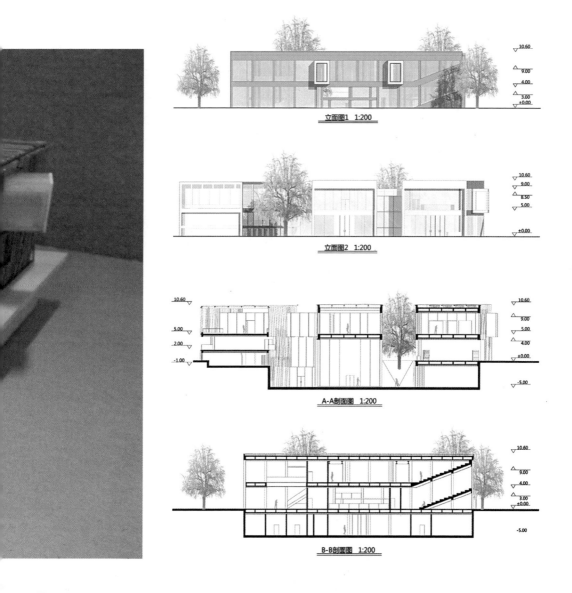

过程分析

空间分析

本页上图：生成过程分析图。**本页下图**：空间分析图。**对页**：构造分析图。

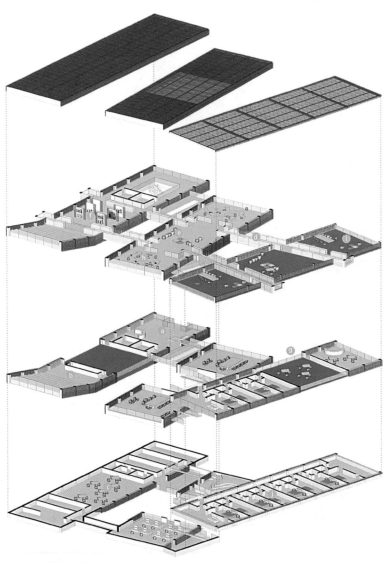

平面功能空间轴测图

The house of architects is located in the Guihe Creative Park in the northwest of Beijing to hold architect camp, integrating the functions of living, working and communication. The people participating in exhibition and exchange are easily guided to the accessible area, initiating more abundant activity and creating free space for architects. The openness of the plot decreases gradually from the northwest to the southeast, and the privacy gradually improves. The good landscape orientation in the southeast side and the sunlight provide comfort and is preferred for the living and leisure. Therefore, the plot is divided into three parts, namely communication, working and living in function. Three functions require intense connection. The horizontal passage attracts people to move towards the communication space and viewing platform, creating a communication space of architects to the outside world in the working and living space. The reed stalks are used as the skin of main buildings. Such building spaces face each other with certain independence. Using local materials is beneficial to saving costs and utilization of local resources, promoting the rational use of materials and protecting the environment.

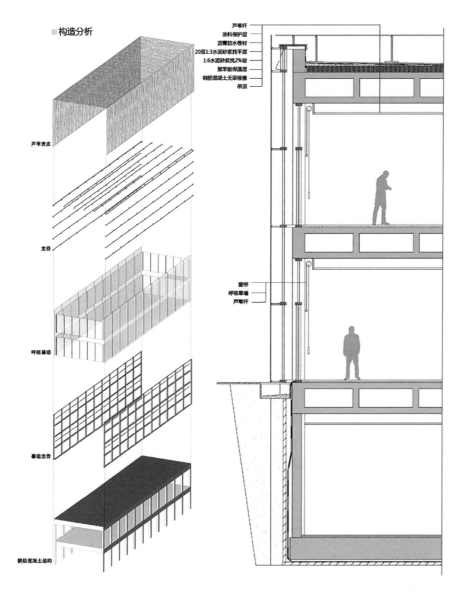

教师点评

孙冉同学是我们课程设计组一位给人印象很深的女同学,她在设计课中表现出女生特有的认真严谨的学习态度。每次课前都做大量的资料收集研究工作,因此课上可以与老师有深入的讨论。她的设计构思来自对建筑基地现场的观察,基地旁的水岸边存在着大量芦苇杆,因此想到用这种天然的材料来构建未来的建筑。孙冉的方案通过三个成不同角度排列的长方形盒子自由地布置建筑功能,建筑表皮饰以芦苇杆构成的肌理,用"对望芦苇荡"的概念来表达设计师工作营设计构想。这种从建筑的技术建构着手深化设计是一种理性的方法,建筑专业的学生应该掌握这种技巧来发展和丰富自己的建筑构想。虽然芦苇杆技术构造的研究并不成熟,但这种回归建筑本质的技术态度是值得鼓励的,也对孙冉同学未来的建筑学业十分有帮助。

Teacher's comments

Sun Ran is a very impressive female student in our design team, and she expresses peculiar careful and rigorous learning attitude of a schoolgirl in the design course. Every time, she will collect massive amount of data for research before course, and support in depth discussion with the teachers in class. Her design concept is from the observation to the construction site. There are a large number of reed rods along the shore beside the base, so this natural material is considered to be used for the construction of future buildings. Through three rectangle boxes arranged in three different angles, scheme of Sun Ran freely arranges the building function. Surface of the building is decorated with textures derivate from reed rods. She expresses the design idea of work camp of the designer with the concept of "LOOK, THE REEDS". Students majoring in architecture shall master the skill of the rational method of design development from the perspective of construction technology, to develop and enrich the architectural concept. Although the research on the structure system of the reed rods is not mature, the technical attitude of returning to essence of architecture deserves encouragement, which is also helpful for Sun Ran's future course of study on architecture.

对页(从上至下):庭院。办公。室外会议。**本页下图:** 客房。

三合院/
COURTYARD

项目选址：北京沙河创意产业园
项目类型：办公、居住
建筑面积：4325m²
用地面积：2956m²

方案设计：刘畅
指导教师：邵韦平
完成时间：2014

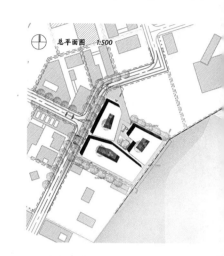

"三合院"是一座"建筑师之家"，在妫河的一块空地进行假想设计。其设计初衷，是考虑到建筑师的工作场地和私人场地原本没有固定联系，我们希望通过"建筑师之家"，让Career & Life在建筑空间中和谐实现。

三合院"建筑师之家"将容纳三个团队或组织（每个团队或组织按8人计）。不同设计团队和组织之间在日常的学习切磋，让Group & Group & Group 的模式以"三合院"方式存在。

其中，建筑的功能空间由三个模块递进组成。Exhibition (PUBLIC) -> Workshop (SEMI-PUBLIC) -> Home (PRIVATE)，形成明显的功能区分和递进关系，从而让人在特定的场所进行特定的表现而不至于含混。另外，在生活模式上，三合院强调，在任一个场所，都可以不断定义私人独处和多人交流的空间。

开篇：主要方案图。**本页上图**：总平面图。**本页下图**：内部庭院。**跨页**：主要方案图。**对页上图**：立面和剖面图。

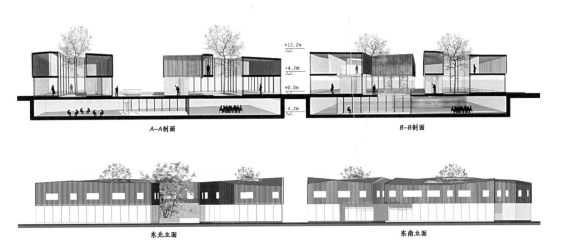

A-A剖面　　　　　　　　　　　B-B剖面

东北立面　　　　　　　　　　　东南立面

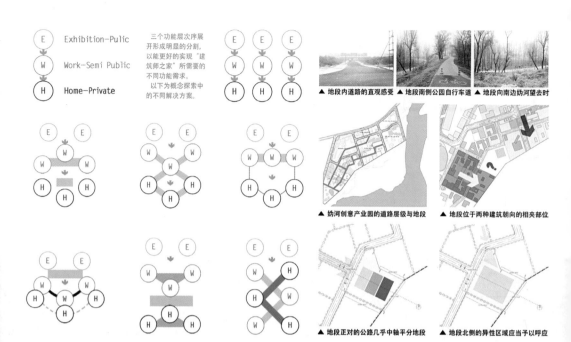

本页上图（从左至右）：概念分析图。地段分析图。本页下图：功能分析图。对页图（从上至下）：形态生成分析图。屋顶生成分析图。三合生成分析图。

外来入口：分别设置，且均预留入口广场；同时内部出入口面向内庭

纵向交通：三个独栋各自沿内庭设楼梯；同时有后勤交通及公共交通

横向交通：沿场地和建筑各自轴线展开横向交通；有室内外两种连廊

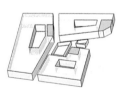

吹拔空间：依靠外来入口设置，丰富空间感受

工作区：位于三合院的"第二进"

讨论区：在首层流线末端设置。

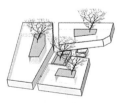

绿化空间：栽有植被及树木，内外相融。

流外部线：方便消防使用以及后勤供给。

三体量形成三合关系　变化以顺应场地边界　引入第二套轴线方向　插入公共下沉庭院　分别考虑入口准备空间　各自三合、深入概念　设置屋顶坡面，调整

在3x3网格中插入控制棒以调节标高　　每个控制棒有0、90、180cm三中标高，按行列错开进行移动　　将生成好的整体屋面套入三个建筑形体，取二者交集　　最终屋面富于变化但不失连贯性，让游者在行进中感受到三者的统一

The "COURTYARD" is a "House of Architects" that is envisioned in a space of Guihe. The original intention of design is to take into account that there is no permanent link between the working place and private space of the architect, and we hope that the Career& Life can be achieved harmoniously through the "House of Architects" in the building space. The COURTYARD "House of Architects" will accommodate three teams or organizations (each team or organization include eight persons). Different design teams and organizations learn from each other in daily work, enabling the inter-group communication to exist in the form of "COURTYARD". The functional space of the building consists of three modules, including Exhibition (PUBLIC) - > Workshop (SEMI-PUBLIC) - > Home (PRIVATE), which form distinct functional distinction and progressive relationship, so that people will not be confused in implementing specific performance in a particular place. And for the mode of life, COURTYARD stresses that the space for private stay and communication between several people can be continuously defined in any place

  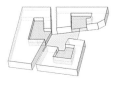 

三个独栋建筑体量首先"三合"，空出一边　　每个独栋在内部形成四合庭院，深入概念　　各个四合院在首层开放一边，与公共场地融合　　三合元素反复在不同层次出现，强化概念

  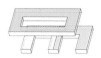 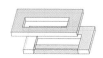

传统合院常有三进：前厅（白）、正厅（黄）、后厅（绿）层层深入　　将用于居住的后厅上移至二层，对传统三进进行概念变形　　进而将用于接待的前厅和正式工作的后厅分别处理为"三合"形式　　对首层空间继续变化，创造整体三合的形体，并通过室内外廊道连接

构造演示

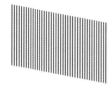

木格栅尺寸：厚3cm，长10cm，间距50cm

木格栅密排后，一方面可以为内部的住宿空间提供私密性保护，一方面不影响内部空间向外的通透性

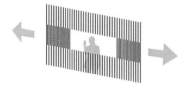

木格栅窗：在1.0m到2.2m为可拉开的窗扇

在需要通风、开敞环境时，拉开格栅，一赏三合院的美景和妫河美丽。

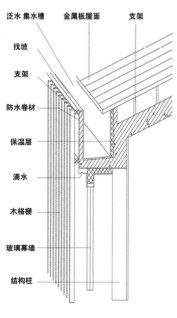

本页左图：构造图。跨页图（从上至下）：二层效果图。内部效果图。

教师点评

三合院并不是三面围合的院子,而是由三个形态各异的四合院构成的建筑组团所形成的设计主题。设计者在一个缺乏场地限定条件的基地上,巧妙地用一个带有地域特征的"合院"概念将环境、功能、文化要素恰当地整合在一起,完成了一个比较优秀的课程设计作业。

总图设计上,在靠近人流方向是一个梯形开放空间,经过它来到中心广场,在这里可以看到三个合院的主入口。三个院落根据各自场地条件自由布置,灵动而有秩序。较大的院子拥有街道和亲水条件,两个小一点的院子分别有其中一个条件。建筑的外观处理的比较完整统一,屋顶的肌理延伸到墙面,接近地面的立面是透明的,让建筑更加开放。

刘畅同学的专业基础很好,很善于沟通,在经过短暂的曲折之后很快找到了感觉,设计越做越顺,达到了课程培养的目的。

Teacher's comments

COURTYARD is not the courtyard with three sides, but a design theme formed by the building groups consisting of three quadrangle courtyard in different shapes. On the basis of lacking specific site condition, the designer skillfully integrates environment, function and culture elements in a proper way by taking advantage of the "courtyard" concept with regional features, and completes an excellent course design work. For the design and the general drawing, an open trapezoid space is placed close to the direction of stream of people, through which you can reach the central square. Here you can see the main entrance of the courtyard. The three courtyards are arranged freely in accordance with respective site conditions, which are flexible and orderly. The bigger courtyard is provided with street and hydrophilic conditions, and the relatively smaller ones are respectively provided with one of these conditions. Treatment of appearance of the building is complete and unified, texture of the roof is extended to the wall, and facade close to the ground is transparent, all of which make the building more open. Liu Chang has a very good foundation of architecture specialty, and she is also good at communication. After transient dilemma, she finds the strategy. The design process becomes smoother, and the purpose of course cultivation is accomplished.

筑·园/ARCHITECT PARK

项目选址：北京妫河创意产业园
功能定位：建筑师之家
建筑面积：3600m²
用地面积：2957m²

方案设计：刘凤逸
指导教师：邵韦平
完成时间：2014

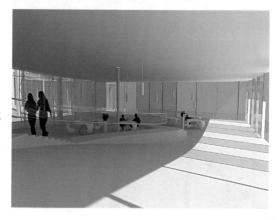

本方案致力于营造纯粹、通透的空间体验，通过起伏的建筑形体自然而然的将人流引入一个内向型的庭院，再引入建筑内部。室内公园式的流线可以让使用者在一个大空间中自由活动，路径可以自己设定。建筑师的工作空间与生活空间的关系根据使用者个体的不同也是若即若离的，形成了立体的、模糊了内外界限和楼层概念空间模式。

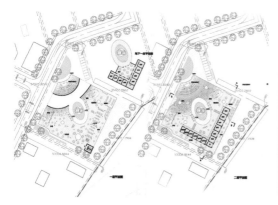

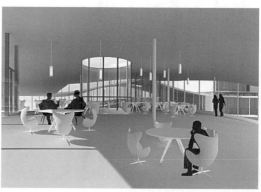

建筑师周末俱乐部/
ARCHITECTS WEEKEND CLUB

项目选址：清华建筑学院南侧花园
项目类型：建筑师周末度假、放松、简单与创意性工作场所
建筑面积：4800m²
用地面积：2956.9m²

方案设计：王之玮
指导教师：邵韦平
完成时间：2014

该设计坐落之地，正适合周末休闲，环境清幽；俯看妫河，自身形态波动曲折，对应水势山形，却又首尾相连，抱成环形，于是人得以自然漫步，却又时时能有机会与人交流，不会显得孤独。开放而静谧，是设计的意趣所在。

建筑师之家/ARCHITECTS' HOME

项目选址：北京妫河创意产业园
功能定位：供建筑师集中居住，健康工作和充分交流的场所
建筑面积：2900m²
用地面积：1700m²

方案设计：康凯
指导教师：邵韦平
完成时间：2014

建筑师之家方案运用生态设计策略，采用ETFE膜气枕作为建筑表皮，充分利用自然采光和自然通风，达到低碳节能的目标。回字形的平面布局和剖面布局辅以充足的室内绿化，为建筑师的工作和交流创造积极的空间体验，面向妫河的玻璃幕墙营造绝佳的景观面。

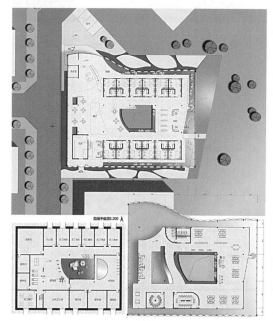

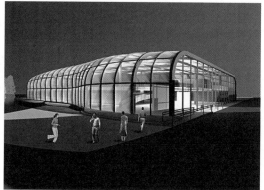

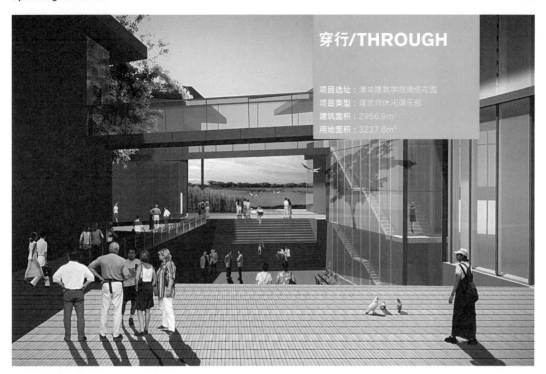

穿行/THROUGH

项目选址：清华建筑学院南侧花园
项目类型：建筑师休闲俱乐部
建筑面积：2956.9m²
用地面积：3237.6m²

方案设计：杨烁
指导教师：邵韦平
完成时间：2014

项目位置位于北京妫河地区，曾为"延庆八景之一"，风光秀丽。建筑师长期于城市纷扰中加班加点，身心疲惫，需要一个放松的场所。在此处建造"建筑师之家"这一项目，是为了在设计任务相对较少之际，增加生活中的休闲成分，让设计师参与唱歌、野餐、冥想等更丰富的活动，回归本源的建筑师，让更丰富的活动激发灵感，这也是建筑创意产业园区的建设初衷所在。

建筑师之家/ARCHITECTS' HOME

项目选址：北京妫河创意产业园
功能定位：建筑师之家
建筑面积：3967m²
用地面积：3574m²

方案设计：段俊毅
指导教师：邵韦平
完成时间：2014

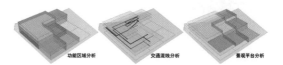

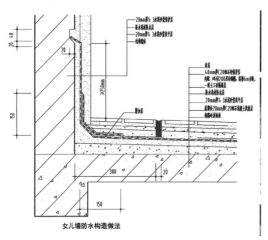

女儿墙防水构造做法

该设计为建筑师之家，考虑到现在建筑师的工作环境，大多在单调的工作隔间中对着电脑工作，生活单调。我设计的建筑师之家是集工作、健身、休闲、娱乐和交流为一体的综合体。工作中间直接有丰富的公共空间，底层敞开对外开放。

建筑师之家/ARCHITECTS' HOME

项目选址：清华建筑学院南侧花园
项目类型：家庭办公 + 公共活动
建筑面积：3260m²
用地面积：2957m²

方案设计：郑松
指导教师：邵韦平
完成时间：2014

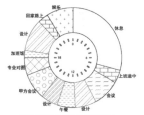

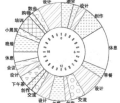

典型的建筑师生活的特点是规律性很强，但生活内容相对单调，生活习惯不够健康

本源生活的内容更加丰富和健康，从建筑师的角度出发兼顾工作、家庭、朋友和个人兴趣

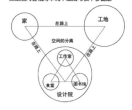

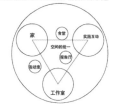

建筑师之家旨在探究建筑师的本源生活，通过对典型建筑师生活的研究和作为建筑系学生的亲身体验，从以人为本的角度分析建筑师的生活需求。典型的建筑师生活形成的原因是空间的分离导致了行为的分解，使得生活趋于单调。建筑师之家将家庭、工作室和实践互动空间集中在一起，创造出灵活的混合功能空间。建筑内集中了满足建筑师工作和生活的多种功能，设计了丰富的流线。建筑底层架空，将地面还给城市，与环境形成良好互动。简洁的形体，精致的结构突出建筑的纯粹性。

建筑师之家/ARCHITECTS' HOME

项目选址：北京幼河创意产业园 01-06-22 地块
功能定位：文化建筑
建筑面积：2893m²
用地面积：2957m²

方案设计：黄河
指导教师：邵韦平
完成时间：2014

该项目定位于为建筑师和相关领域的艺术家提供工作和交流的场所。作者认为交流合作是设计与创造的基础。针对这类传承手工作坊匠人团体工作的特点，本方案借鉴传统聚落的空间模式，塑造从不同层次促进个人、团队、公众的交流的空间。同时建筑内部保留自由度，让建筑里每天发生的故事由身处其中建筑师的工作生活来定义。

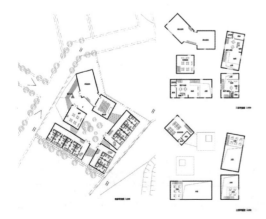

隐·居/HIDDEN HOME

项目选址：清华建筑学院南侧花园
项目类型：建筑师
建筑面积：3400m²
用地面积：1700m²

方案设计：徐珂
指导教师：邵韦平
完成时间：2014

古代文人雅士意义上的"隐居"多有隐匿于大野山林之中，然后修身养性，享受田园生活之意。这里的"隐居"取其拆词之后的涵义，将"隐"字归结为与大自然最大程度的接触，对建筑师而言，就是短暂脱离城市的喧嚣，来到宁静的隐居之所，或设计，或度假；将"居"字对应为居家办公、居家设计之意，旨在给设计师一个更加宽松的氛围。

对应到建筑形式上，通过覆土建筑这一形式达到增加和自然的接触以及将建筑隐匿于自然之下的目的。同时，让建筑主要的一面朝向主要景观，增加对景的应用。并在适当的地方增加庭院，同时将东南方暂时的道路改为庭院绿地，使得人们可以在庭院徜徉，漫步至湿地花园，再漫步至美丽的妫河河畔——荻花之野。

底部的居家办公场所，通过与庭院地面形成700mm的高差，形成如图所示的桌面与草地相连的奇妙景观，居家办公将为设计师提供大面积的绘图平台，而不单单是现代化的办公方式。通过回归最原始的设计方式，使设计师获得更为广阔的思维。居家办公的场所提供了自由的视野和高度，鼓励设计师们回归本源，通过涂鸦和绘画来实现灵感的迸发。

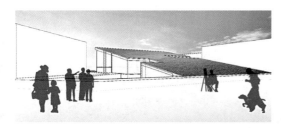

通过限制创造

胡越

北京市建筑设计研究院有限公司 胡越工作室
主持建筑师、公司总建筑师、全国勘察设计大师

"无限"的可能

胡越
北京市建筑设计研究院有限公司 胡越工作室
主持建筑师、公司总建筑师、全国勘察设计大师

教育背景
1982年－1986年
北京建筑工程学院 建筑学学士
2006年－2011年
清华大学 建筑学博士

工作经历
1986年至今
北京市建筑设计研究院有限公司

主要论著
《塑料外衣》[M].同济大学出版社，2016
《建筑设计流程的转变》[M].中国建筑工业出版社，2012
《定制设计的困境》[J].建筑学报，2011/3
《建筑设计方法变革的哲学思考》[J].建筑评论，天津大学出版社，2012
《设计流程结构的应用》[J].建筑学报，2008/12
《大型体育馆的三种模式》[J].建筑学报，2008/7
《几何游戏——望京科技园二期设计》[J].时代建筑，2005/11
《方案竞赛制度的现状及思考》[J].规划师，2000/5
《从衣服到皮肤——一种新型玻璃幕墙》[J].世界建筑，1999/4
《联想的联想》[J].建筑创作，1999/2
《窗式幕墙的设计与实践》[J].建筑创作，1998/2

设计获奖
2000 － 北京国际金融大厦获全国第九届优秀工程设计金奖
2004 － 秦皇岛体育馆获全国第十一届优秀工程设计银质奖
2008 － 望京科技园二期获全国第十二届优秀工程设计银奖
2009 － 五棵松体育馆获获2008年度全国优秀工程设计金奖
2009 － 上海青浦体育馆训练馆改造获北京市第十四届优秀工程评优三等奖
2011 － 上海世博UBPA办公楼获北京市第十五届优秀工程设计二等奖
2013 － 北京建筑工程学院大兴新校区6号综合服务楼获得中国建筑设计奖（建筑创作）金奖

代表作品
北京国际金融大厦（图1）、北京建筑工程学院大兴新校区6号综合服务楼（图2）、望京科技园二期（图3）、上海青浦体育馆训练馆改造（图4）、上海2010世博会UBPA办公楼（图5）、五棵松体育馆（图6）

通过限制创造"无限"的可能

三年级建筑设计（6）设计任务书
指导教师：胡越

课程描述
学习一个有用的设计方法；学习在限制条件下生发设计创意；本课通过设计一个位于创意园区内的小型建筑，向同学介绍一种行之有效的设计方法。学生通过课程学习，初步了解建筑方案的设计过程，学会如何分析设计任务书、形成设计师任务书，如何寻找方案构思的出发点，如何将构思转化成造型语言，如何组织几何元素，在方案细化过程还会涉及到总图、交通流线、功能分析等内容。

课程目标
目标1: 通过阅读任务书和了解建筑所处的环境，发现独特的问题
目标2: 提出设计要点并形成设计师任务书
目标3: 了解设计构想的来源和类型
目标4: 寻找几何原型并与设计要点发生联系
目标5: 了解几何操作的基本方法

项目概述
项目名称： 北京妫河•建筑创意区创意工作室示范街区
项目位置： 北京市延庆县延庆镇西屯村西南地块北京妫河•建筑创意区，距延庆县城2km。
设计范围： 本次设计范围为园区内 01-06 地块（建筑创意区创意工作室示范街区）D 区，规划北三路以南、规划西一路以西的局部区域范围内，具体位置参见园区平面图。

现状及规划设计条件
区位关系： 本项目位于北京延庆县延庆镇西屯村西南地块北京妫河•建筑创意区 01-06 地块内，具体位置详见地块区位。
现状地形： 边界: 地块北侧为规划北三路,东侧为规划西一路(01-06 地块总平面图)。2.2.2 高程（见 01-06 地块总平面图）
现状绿化： 已基本完成园区道路和园区绿化施工，并部保留有原大树。2.2.4 用地规划技术经济指标
用地性质： 教育科研设计用地：总占地面积:8616m² 楼座占地面积:2261m² 地上总建筑面积: 6783.9m² 建筑控制高度: 12m 容积率:0.8
建筑密度： 30% 绿地率:40% 空地率:70%

建筑定位及设计要求
北京妫河•建筑创意区遵循延庆总体规划的要求，作为北京延庆设计创意产业园区一期项目，启动并引领地

区创意产业发展；打造国际一流集创作、培训、科研、成果展示、文化交流等全面功能的创新型文化创意产业聚落；形成北京市泛建筑创意产业及相关行业交流的最重要主流平台。依托延庆县的环境资源优势及其妫水河城市生态景观带，倡导生态、环保、节能、科技、创新等新理念，主张贯穿全程的绿色设计、绿色建筑、绿色运营，北京妫河•建筑创意区将成为创新型生态环保园区的典范。示范街区作为园区重要组成部分将引领园区建设进程，街区中间设东西贯通的公共活动空间，建筑物单体按独栋创意工作室设计，考虑个性化设计的同时，又要求单体设计与园区整体的设计定位和风格相呼应，旨在为设计师提供具有环境、功能、空间、生态和文化吸引力个性化独立创作空间，使设计师和艺术家在创作时既能享受生态创作空间和创作环境又能激发创作灵感。建筑物请按街区"规划条件"（pdf 文件）要求布置。考虑到市场的需求，建筑物应考虑可分成 2~3 个单元，每个单元应能独立使用，也可合并 使用。建筑物需设有休息室、起居室、工作室、展室、书房、健身房、服务用房、茶水间(满足基本餐饮需求)、储藏室、基础设施用房(单独计量且室外设独立入口)、街区中间公共活动空间的地下一层为公共车库(为每栋建筑物提供 4~6 个车位)，每栋单体建筑需设置与地下公共车库之间连接通道。

功能建议： 建议在建筑物一层设置作品展示空间和独立庭院花园，用于设计师接待来访者并与之进行交流沟通，创意设计工作室建议设置在二层，三层主要用于休息和其他配套用房。

建筑面积(地上)： D栋 1173m^2(建筑面积可适当增加，但须控制在 1300m^2 以内)

造价控制目标
地上建安成本(不含室内装饰): 3500元/m^2;
地下建筑建安成本: 3500元/m2m^2。

材料和设备选用要求
按照成本控制原则，采用技术成熟、性价比高的材料、设备等，保证项目的高品质并符合绿色建筑设计标准并充分考虑延庆气候情况。延庆位于北京西北方向，气候较北京寒冷，各专业设计应充分考虑当地的地理及气候特征，并应对冬季使用问题予以充分关注。

设计目标
提出和谐、完整、特色的规划。根据城市规划设计要求，确定建筑、空间、道路广场、景观等方面的整体特色，提出建筑控制、空间控制、景观控制、环境保护、生态维护等方面的文件，使建筑物与庭院有机结合。希望建筑师能提供室、内外空间设计的统筹协调的设计文件，创造出崭新、积极、强烈的新型空间特征。示范街区的单体建筑设计将在总规划师的指导下进行，各栋单体建筑设计可采用不同风格、不同流派的建筑创意汇聚构成整体的特色建筑体验，确保地块整体的建筑形象丰富而和谐。

艺术家工作室 /
ARTIST STUDIO

项目选址：北京纺河创意产业园
项目类型：居住、工作室
建筑面积：1357m²
用地面积：1108m²

方案设计：甘旭东
指导教师：胡越
完成时间：2014

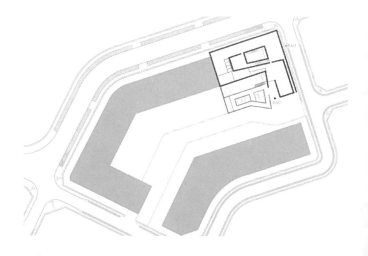

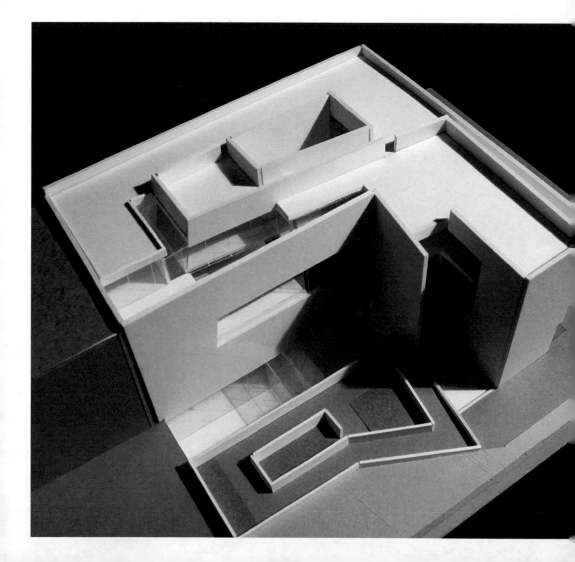

开篇：主要方案图。**对页上图**：总平面图。**跨页图**：主要方案图。**本页中图**（从上至下）：一层平面图。二层平面图。三层平面图。**本页右图**（从上至下）：南立面图。北立面图。建筑内部关系。

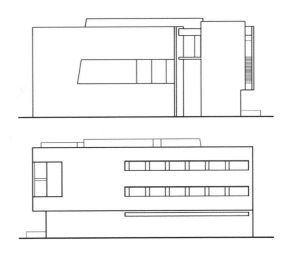

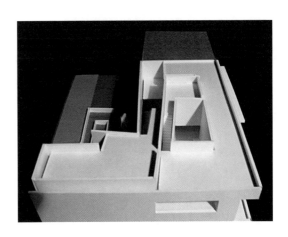

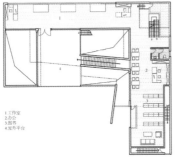

地处妫河创意文化园区的建筑受到规划条件的限制，层高，贴线率等都有严格的要求，导致建筑外形的固定。建筑从艺术家和来访人之间的关系模式的思考出发，按照艺术家工作室服务对象的亲疏关系将人分为三个深度不同的层次，也就是最远层次——陌生游人，中间层次——来访熟客以及核心层次——个人家庭。他们所需求的建筑功能的深度也不同，将这样分层的关系映射到场地之上，形成了分层的空间，同时一条由外部螺旋向内的流线串联起从公共到私密的建筑各部分功能，使得各层群体对不同深度功能的需求在这条渐进式的流线上都可以得到满足。由最初的螺旋线生成了墙体这一几何原型，限定出了旋转向内的多层空间。通过对这一单一墙体进行偏移，翻折等操作以达到实现引导人旋转向内的目的。同时平面功能的布置也是在这一条流线的基础上进行排布的。

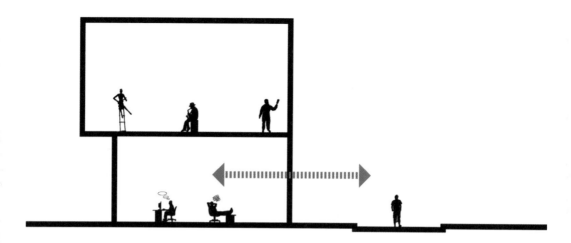

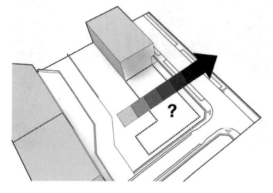

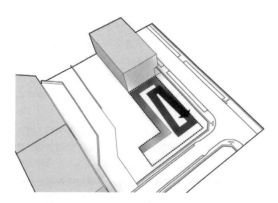

The building located in Guihe Cultural Creative Park has been restricted by the planning conditions, the floor to floor height and the set-backs, leading to the final massing of building. Starting from the relationship between the artist and the visitors, three different depth of floors are divided for people according to the affinities of artists and their clients. That is the farthest level—strange visitors, intermediate level—frequent visitors and core level-- individual households. Their demand for the depth of architectural is also different. The hierarchical space is formed when the hierarchical relationship is mapped onto the site. Meanwhile, an streamline spirals inwards from the external side connects the functions from public to private, making that the demands for different depth of functions of different groups can be satisfied in the progressive streamline. The geometric prototype of the wall is generated from the original spiral line, which defines the multi-layer space that spirals inward. Through the offset, turning and other operations of this single wall, the purpose of guiding people to turn inward is achieved. At the same time, the layout of plane functions is also arranged based on this line.

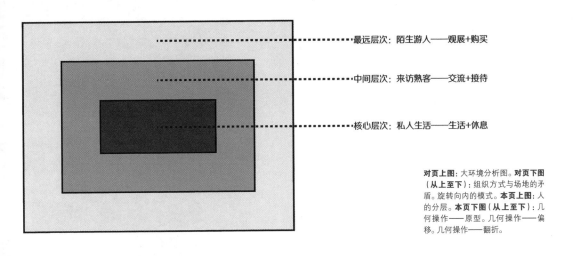

最远层次：陌生游人——观展+购买

中间层次：来访熟客——交流+接待

核心层次：私人生活——生活+休息

对页上图：大环境分析图。**对页下图（从上至下）**：组织方式与场地的矛盾。旋转向内的模式。**本页上图**：人的分层。**本页下图（从上至下）**：几何操作——原型。几何操作——偏移。几何操作——翻折。

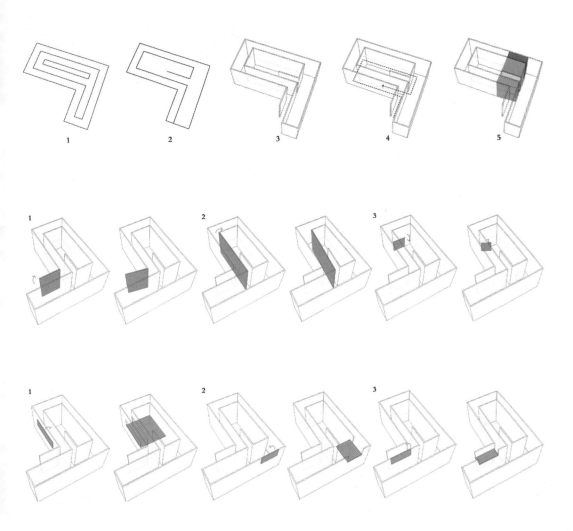

跨页图（**从上至下**）：主入口透视。三层平台透视。**本页上图**：庭院透视。**本页下图**：二层平台透视。

教师点评

这份作业很好地达到了本课程对学生的基本要求。学生在任务书解读、设计要点转化成形式语言以及几何操作等方面表现出了较强的能力。学生在设计螺旋上升的主要空间时，能够结合院落、开窗、平台等元素，表达一种与设计主题相契合的分为，说明该同学已经对建筑学的意义有较为深刻的领悟。

Teacher's comments

This work well meets the basic course requirements for students. The student shows strong capacity in interpreting assignment book, transforming key points of design to form language, and geometric operation. In designing main spiralspace, the student can express a kind of atmosphere meeting the design theme by combining elements of courtyard, window, platform, etc. This indicates that this student has deeply understood the significance of architecture.

视界/HORIZON

项目选址：妫河创意园区示范街区 D 地块
功能定位：创意园区艺术家会所设计
建筑面积：1020m²　　**用地面积**：1430m²
居住（地下）：470m²　　**聚会**：370m²
展览：650m²　　**室外广场**：430m²

方案设计：孙鹏程
指导教师：胡越
完成时间：2014

开篇：主要方案图。**本页及对页**：分析图。

设计概念
——灵感催化剂

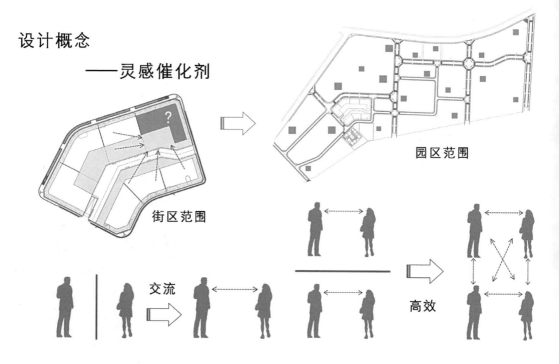

街区范围　　园区范围　　交流　　高效

空间几何操作

STEP1　在限定的空间范围内生成单体空间的控制点

STEP2　根据VORONOI算法生成空间单体

STEP3　对空间单体编号

STEP4　移除底层的部分空间单体形成一个开放的聚会交流空间

STEP5　剩下的空间单体形成包裹聚会空间的外壳，作为展览空间

STEP6　打破空间单体边界，互相交流

STEP7　选定正方形作为开洞的形态，角度和大小作为参数，具有景框感

STEP8　设置一条贯通的通道在展览空间单体中穿行

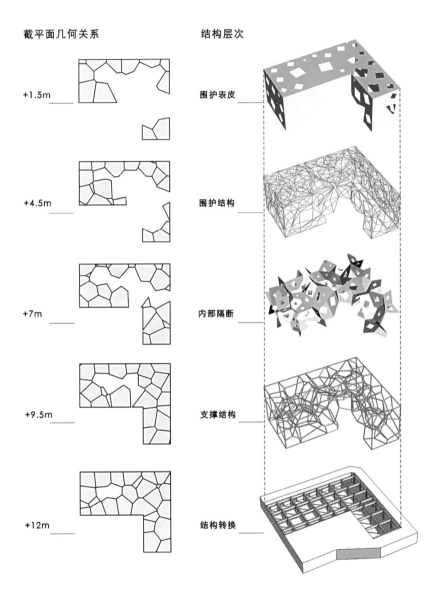

设计为妫河创意园区示范街区D地块的建筑单体设计，主题是"有限的条件创造无限的可能"。"有限的条件"指的是建筑场地红线、控高等很多硬性控制条件，"无限的可能"指的是建筑的功能和氛围。

本设计考虑了街区和园区不同尺度的需求，将甲方设定为一名艺术策展人，在创意园区中为创作者们提供思想交流的场所。艺术来自灵感，交流和启迪是灵感的重要来源，艺术家们在此邂逅、交流、构思、冥想、观展……本设计希望建筑打破传统的空间结构，空间结构应该适合艺术家们的高效连接。

本页：总平面图。对页：平面图、立面图及剖面图。

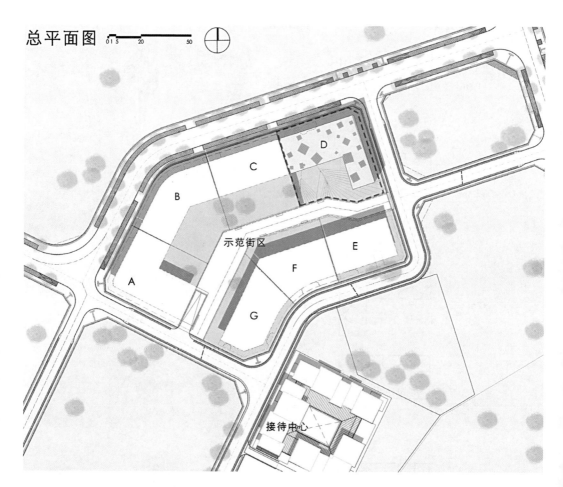

总平面图

The design is involved with the monomer building design in Plot D, defining Street, Guihe Creative Park, with the theme of "limited conditions to create infinite possibilities". "Limited conditions" refer to the red line of construction site, height control and other restrictions from building codes, and "infinite possibility" refers to the function and atmosphere of the building.

This design takes into account the needs of different scales of street blocks and parks, Party A is set as an art curator who provides the place for the creators in the creative park to exchange ideas. Art comes from inspiration. Exchange and enlightenment is an important source of inspiration. Artists encounters each other here for communication, thinking, meditating and exhibition viewing... We hope that the building can break the traditional spatial structure, and the spatial structure should be suitable for the efficiency of artist's connection.

+1.6m标高层平面图 1:200

-3m标高层平面图 1:200

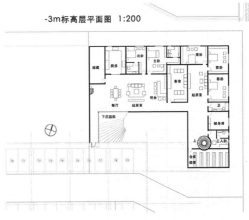

+13m标高层平面图 1:200

+7m标高层平面图 1:200

东立面图 1:200

南立面图 1:200

A-A剖面图 1:200

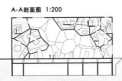

B-B剖面图 1:200

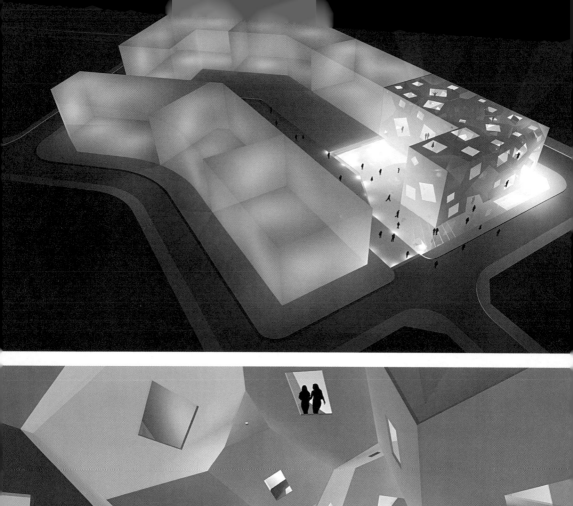

本页及对页: 效果图。

教师点评

这份作业是小组中唯一一份用参数化手法进行设计的作业。借助工具进行几何操作是建筑设计的一个重要的方法，将工具的特点与设计的要求和建筑的内在逻辑协调好是设计成败的关键。该设计在设计要求与工具的局限性之间取得了平衡。作者借助参数化设计营造的多样的内部空间与业主的职业特点吻合得很好。

Teacher's comments

This workis the only one designed with parametric method in the team. Geometric operation by virtue of tools is an important method of building design. The key to the success of design is the proper coordination of characteristics of tools, design requirements and internal logic of the building. This design balances the design requirements and tool limitation. The diverse inner spaces created by the designer using parametric design are properly identical with vocational characteristics of the owner.

根雕艺术家工作室 / CARVING ARTIST STUDIO

项目选址：北京妨河创意产业园
功能定位：创意园区艺术家个人工作室设计
建筑面积：1357m²
用地面积：1108m²

方案设计：李旻华
指导教师：胡越
完成时间：2014

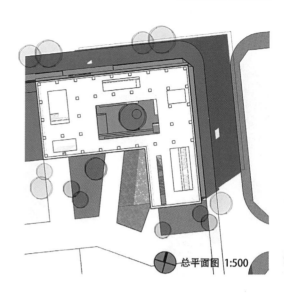

总平面图 1:500

开篇：主要方案图。**本页上图**：总平面图。**本页下图**：主要方案图。**对页**：剖面图及立面图。

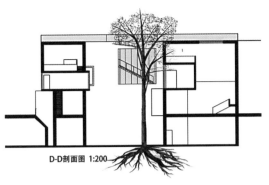

D-D剖面图 1:200

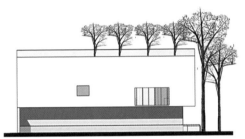

东立面图 1:200

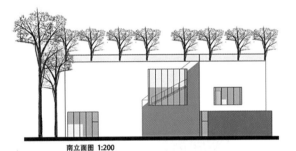

南立面图 1:200

设计要点

大环境：地块自然环境良好，建筑设计充分利用自然的景观河资源。

小环境：整个艺术家工作室示范街区的屋顶为连续开放空间，D地块位于东北边缘，作为这一屋顶连续空间的起点——平坦的屋顶大平台和景观。

功能

艺术家艺术创作和个人生活两种空间分别成为独立体系，在一个特殊空间两者相遇。

对象应对

根雕艺术创作客观条件要求：室内工作间避光干燥，室外操作平台根雕艺术品与建筑本身相融，"无处不在"重点营造——艺术创作空间与个人生活空间相遇的中央核心筒从地下贯穿屋顶的垂直"深井"（地下工作室的室外工作平台）。

六个水平向插入的盒子。

盒子成为连通建筑外轮廓和中央核心筒的"桥梁"，整个大空间减去中央核心筒和盒子，得到的负型形成了丰富的展览空间。

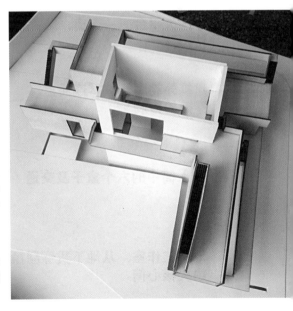

生成过程——立体几何操作

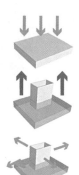

根据根雕艺术家的工作特点,将其主要工作空间全部置于地下

产生一个从地下贯穿屋顶的室外庭院沟通地上空间,使艺术家以及参观者艺术与生活"兼得"

垂直庭院为设计重点营造的空间,所有生活与艺术的交集全部发生在这里

生成过程——立体几何操作

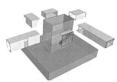

在不同层高向中央垂直庭院中插入个水平向的盒子,盒子内的空间用生活休闲使用

每一个水平盒子均连接中央垂直和建筑轮廓外墙,如同插入建筑实内部的空管子,在生活的各个角落将地上盒子生活—地下艺术工作大间紧密结合

空间拆分——通过几何操作将空间分类

屋顶大平台——天窗、"井口"、枯树阵

建筑外轮廓及被"连通盒"穿破的窗洞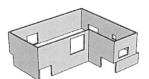

插入筒中的六个盒子及交通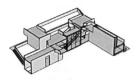

地下工作室、从地下贯穿屋顶的中央核心筒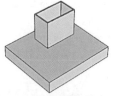

几何操作——平面

平面插入盒子分布

插入盒子两端分别连接建筑外轮廓和中央核心筒——穿破的地方形成门窗

几何操作——平面

交通空间（楼梯及走道）沿着建筑外轮廓和中央核心筒分布

盒子分布有特定功能的生活空间，家具按控制线摆布

几何操作——立面

盒子穿透建筑外轮廓和中心核心筒，立面窗洞由此形成

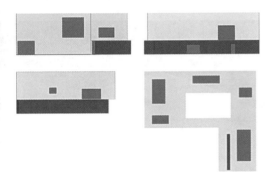

几何操作——立面

盒子穿透建筑外轮廓和中心核心筒，立面窗洞由此形成

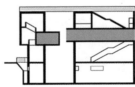

Design features

General environment: The natural environment of this land is good; the architectural design takes full advantage of the natural river landscape and local resources

Local environment: The roof of the demonstration block ofentire artists' studio is a continuous and open space. Plot D is located on the northeast edge, which is used as the starting point of a continuous flat roof platform and landscape.

Functions

The creative and personal life of artists are independent systems that meet in a special space.

Object response

Objective conditions of root art creation: Keep dark and dry in indoor working room, and outdoor operation requires platform

The root carving art and architecture itself are in harmony, existing everywhere.

Focus on creation—the central core meet the artistic creation space and personal living space through the vertical "deep well" from the underground (outdoor work platform of underground studio) Boxes inserted from six horizontal directions The boxes become "bridge" connecting the building facade and central core, the negative-space achieved by reducing the central core tube and box from the entire large space forms rich exhibition space.

本页上图：模型鸟瞰图。**跨页图**：模型主入口半鸟瞰。**对页上图**：室内空间展示。**对页下图**：室内楼梯。

教师点评

我对这个项目的场地进行了严格的限制,建筑体型也被限定在一个L型的方形空间内,这份作业在严格限制的体型内通过院落、大空间内离散的小型封闭空间和平台的摆设,错落的楼梯和结构墙营造出一个丰富的,张弛有度的内部空间。

Teacher's comments

I strictly restrict the site of this project, and also limit the building shape into an L-shaped square space. In the strictly restricted shape, this workcreates an abundant and balanced inner space through courtyard, scattered small enclosed space and arrangement on the platform in the big space, well-arranged stairs and structures.

MW版画家工作室 / VERSION OF THE ARTIST STUDIO

项目选址：汾河创意园区示范街区
功能定位：艺术家工作室
总建筑面积：1003m²　　基地面积：1108m²
容积率：0.9　　　　　　绿化面积：430m²
建筑高度：12m　　　　　层数：4（包括地下室）

方案设计：陈羚琪
指导教师：胡越
完成时间：2014

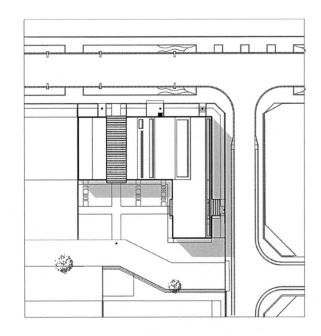

开篇：主入口效果图。**本页上图**：总平面图。**本页下图及对页图**：设计分析。

在这个方案开始前，我从现实中选择了一位朋友作为我这次方案的业主（汤铭伟）。针对这一位业主，设计了这一所版画家工作室。首先，我对他的性格进行了一些总结：孤僻、心思多变、情绪化、压抑、心思细腻、保守、内向、对艺术充满热情等，我还把他以往的作品都仔细欣赏了一遍。其次，我对地段进行分析，希望留有通道可以将人流引入地段中心，增加地段的开放性。

结合地段的分析以及对业主的了解，我将建筑纵向分成了5个区域，一个是私人空间，一个是命名为"木"的通道，另外三个分别是命名为"石"、"草"和"水"的展区，并配合业主作品在这三个展区中设计了5个场景。

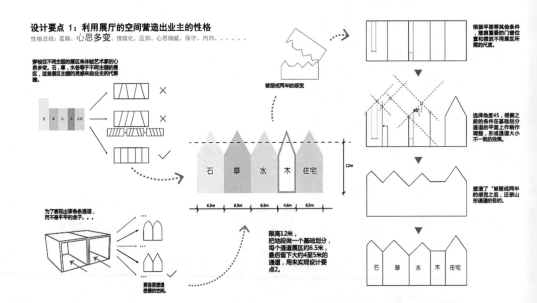

设计要点 2：环境应对之引入人流到步行街

为了增加人流，打破被建筑围死的步行街。增加外围与内步行街之间的交流。

地段：妫河创意园区示范街区D地段
基地面积：1108平方米
退线内基地面积：546
总建筑面积：1218.8
庭院面积：562
停车位：4

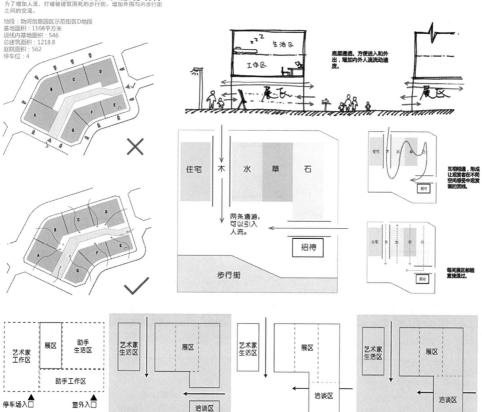

设计要点 3：简化版画制作过程中各种机器使用的流线

8 机器 + 2 其他
美院机器因为不定期增加，加上机器不易移动，所以机器的摆放并不方便复杂流线的丝网印刷制作过程。

丝网版画使用机器的流程：
（分成 3 部分）

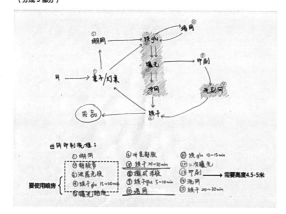

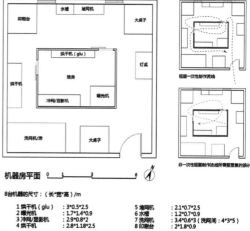

8台机器的尺寸：（长*宽*高）/m

1 烘干机 (glu)　: 3*0.5*2.5
2 曝光机　　　: 1.7*1.4*0.9
3 冲网/显影机　: 2.9*0.8*2
4 烘干机　　　: 2.8*1.18*2.5
5 地网机　　　: 2.1*0.7*2.5
6 水槽　　　　: 1.2*0.7*0.9
7 洗网机　　　: 3.4*0.6*3（洗网间：4*3*5）
8 印刷台　　　: 2*1.8*0.9

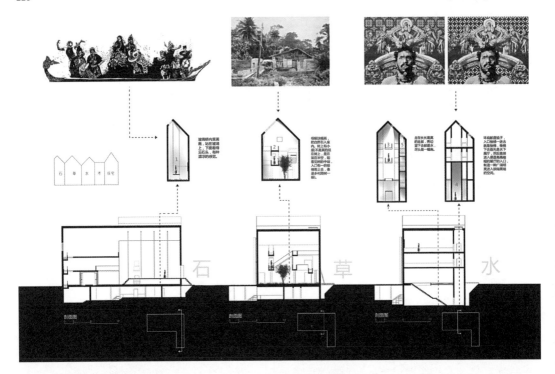

Before this program, I selected one of my friends to serve as the owner of this program (Tang Mingwei), to whom I designed a woodcutter studio. First, I concluded his character: being solitary, varied in thought, emotional, depressed, attentive, conservative, introvertive and enthusiastic in arts, etc. I also appreciated all his former works carefully. Secondly, I analyzed the section which as I hoped could be designed with a passageway leading passenger flow to the center of section, increasing its openness.

Combining my analysis on the section and understanding to the owner, I divided the building into five areas vertically, including a private space, a passageway named "Wood" and three exhibition areas named "Stone", "Grass" and "Water", respectively. Besides, I designed five scenes in these three exhibitions in combination of the owner's works.

对页上图：入口效果图。**对页下图**：通道效果图。**本页图**：室内效果图。

教师点评

这份作业的突出特点是设计人对预先设定的业主有充分的沟通和了解，在深入了解业主的工作状态和作品后，将空间营造出的氛围与业主作品的内在气质结合在一起，作业在完成基本要求的情况下，能够有意识地将设计的气质与设计的主题结合起来是很难得的。

Teacher's comments

Highlight of this work is that the designer fully communicates with and understands the preset owner. After deeply understanding the working status and works of the owner, the designer combines the atmosphere created by the space with the inner temperament of works of the owner. On the basis of meeting the basic requirements, it is rare to consciously combine temperament of design with theme of design.

浮岛/FLOATING ISLAND

项目选址：北京市劝河文化园
功能定位：艺术家工作室
建筑面积：1300m²
占地面积：500m²

方案设计：张启亮
指导教师：胡越
完成时间：2014

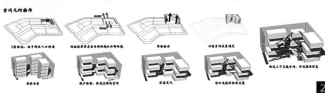

艺术家在生活和工作的状态中，犹如在艺术的海洋中探索。因此设计将整体空间划分为独立的建筑体量，取名为"浮岛"，连接这些盒子空间的楼梯也提供了参观者与艺术家互动的不同视角。

-4m标高平面 1:200

-1m标高平面 1:200

2m标高平面 1:200

5m标高平面 1:200

8m标高平面 1:200

11m标高平面 1:200

地平线/HORIZON

项目选址：北京市延庆县延庆镇西屯村西南地块北京妫河·建筑创意区，D区地块
项目类型：艺术家工作室
建筑面积：1872m²
用地面积：1108m²

方案设计：崔永
指导教师：胡越
完成时间：2014

本项目位于延庆县延庆镇西屯村西南地块北京妫河·建筑创意区D区，场地视野良好，可以看到连续平整的地平线，以此为概念，创造一个通透轻盈的艺术家工作坊。结构与装饰尽可能使用了玻璃这种透明材料，仅以少量私密空间做不可视围合，营造了一种缥缈奇妙的空间体验。

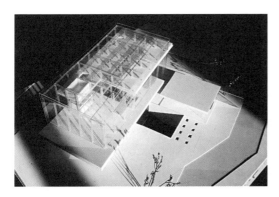

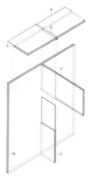

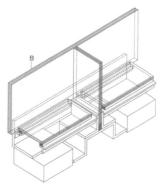

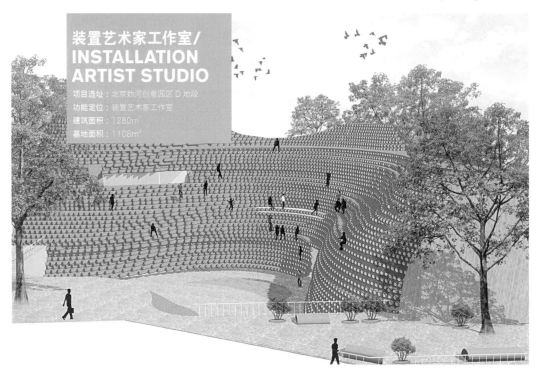

装置艺术家工作室/
INSTALLATION ARTIST STUDIO

项目选址：北京妫河创意园区 D 地段
功能定位：装置艺术家工作室
建筑面积：1280m²
基地面积：1108m²

方案设计：吕代越
指导教师：胡越
完成时间：2014

SPACE EFFECT

本设计在地段限制要求极为苛刻的情况下，从周围环境特征和参观者参观行为分析出发，以周边山岭为空间原型、以周边产业生产的水泥管为基本构件，以在外部塑造围合出室外展场、可供游客攀爬参观的空间看台，在内部塑造可供艺术家进行创作和作品展览的特色大空间为设计要点，希望塑造一个满足使用者需求、富于故事性、戏剧性、空间感和形式感的建筑空间。

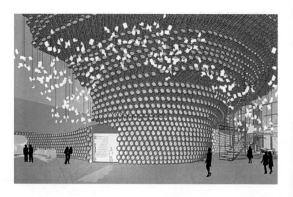

夫妻工作室/
HUSBAND AND WIFE STUDIO

项目选址：北京坊河创意园区 D 地段
项目类型：夫妻共用，家庭工作室
建筑面积：1200m²
D 地段基地面积：1108m²

方案设计：马步青
指导教师：胡越
完成时间：2014

设计课主题是"通过限制创造"无限"的可能"，提出设计要点，组织几何元素和几何操作。设计内容为夫妻艺术家工作室，要兼顾夫妻生活和艺术家工作两种状态间的分离和重合，通过一定的几何原型的提取和操作变化，使得建筑内容达到对内分区明显，对外内向安静的目的。

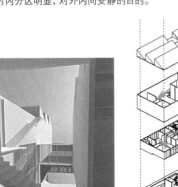

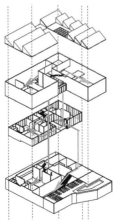

空中温室 / AIR GREENHOUSE

项目选址：北京延庆妫河创意文化产业园
项目类型：艺术家工作室
建筑面积：1120m²
用地面积：1084m²

方案设计：齐大勇
指导教师：胡越
完成时间：2014

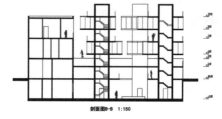

剖面图B-B 1:150

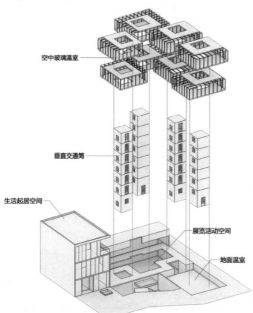

空中玻璃温室
垂直交通筒
生活起居空间
展览活动空间
地面温室

该项目定位于为建筑师和相关领域的艺术家提供工作和交流的场所。作者认为交流合作是设计与创造的基础。针对这类传承手工作坊匠人团体工作的特点，本方案借鉴传统聚落的空间模式，塑造从不同层次促进个人、团队、公众交流的空间。同时建筑内部保留自由度，让建筑里每天发生的故事由身处其中建筑师的工作生活来定义。

雕塑之家 / SCULPTURE HOUSE

项目选址：北京 奶河 · 建筑创意区
项目类型：艺术家工作室
建筑面积：1218.8m²
庭院面积：562m²
地块容积率：1.1
建筑层数：3

方案设计：谯锦鹏
指导教师：胡越
完成时间：2014

在用地红线、高度控制等限制条件下为雕塑艺术家创造一个既能满足其自身大型雕塑创作的场地，也能为游人提供欣赏艺术家创作过程的空间。设计以"墙体"为主题，围绕场地边缘形成半围合的室外庭院，将场地中心让出作为艺术家的室外展场和工作场地。

一层平面 比例1:200

二层平面 比例1:200

地下室 比例1:200

三层平面 比例1:200

艺术孵化器/ART INCUBATOR

项目选址：北京妫河创意产业园01-06-22地块
建筑面积：1300m²
建筑层数：地上3层 地下1层
建筑高度：12m

方案设计：谭婧玮
指导教师：胡越
完成时间：2014

南立面图 1:200

北立面图 1:200

东立面图 1:200

致力于为集群艺术家提供生活保障和公共服务，针对地段环境和对宋庄等艺术区生活情况的实地调研，了解其现状及需求，在设计中进行平衡，形成一个能自我饱和的微型艺术聚落。方案主要从生存、经济、社交、生态四个方面策划，希望能为边缘艺术家提供基本的生活保障，多种形式的买卖平台，多元社交的公共服务，良好的生态环境。整体围绕如何让一个单体建筑形成一个艺术聚落进行设计，致力于以包容、开放的心态向更多的人提供服务，为更多艺术家提供公共空间，希望能作为一次尝试解决现代艺术产业问题的实验。

光影艺术家工作室 / SHADOW ARTIST STUDIO

项目选址：北京市延庆县延庆镇西屯村西南地块北京妫河·建筑创意区
项目类型：艺术家工作室
建筑面积：6784m²
用地面积：8616m²

方案设计：王靖淞
指导教师：胡越
完成时间：2014

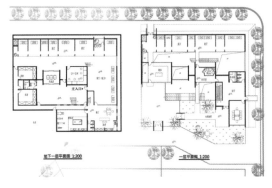

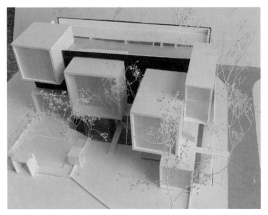

方案以光影画艺术家为虚拟甲方，以光影画艺术家的工作方式与作品的表现形式为设计要点进行设计。建筑设计的重点在于，如何在建筑外边界被严格控制的情况下，找到合理的空间布局与空间互动，使得建筑充满更多的可能。

有限空间的无限组合/INFINITE COMBINATION OF FINITE SPACE

项目选址：北京市延庆县延庆镇西屯村西南地块北京妫河 · 建筑创意区 01-06 地块
功项目类型：文艺术家工作室
建筑面积：6783.9m²
用地面积：8616m²

方案设计：吴珩
指导教师：胡越
完成时间：2014

该方案通过在大空间中置入可灵活组装的空间围合模块，适应雕塑艺术家个性化的空间需求。建筑体量通过任务书限制经过简单几何操作形成，强调整体性和雕塑感，同时形成艺术家创作的主要空间，而小空间则是通过相同模块单元自由围合形成，满足灵活多变的空间需求。

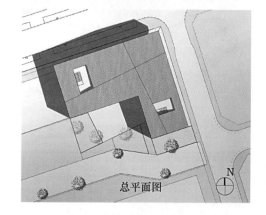

总平面图

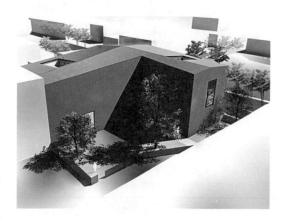

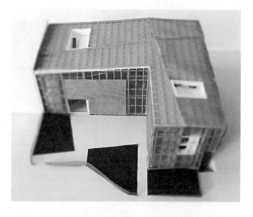

艺术家工作室/ARTIST STUDIO

项目选址：北京妫河建筑创意工作室示范街区
项目类型：艺术家工作室设计
建筑面积：1196m²   **用地面积**：1108m²
容积率：1.08   **绿化率**：0.25%
建筑限高：12m   **建筑层数**：地下1层+地上3层

方案设计：杨茜
指导教师：胡越
完成时间：2014

方案拟定服装设计师和雕塑家两个甲方，营造一种"相遇空间"，室内游人与外街道游人的相遇通过沿街立面的狭长L型空间实现，两个艺术家的相遇围绕中间庭院实现。

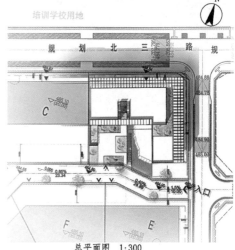

总平面图 1:300

日常经验·个体

梁井宇

场域建筑工作室 创始人主持建筑师

记忆 + 居住畅想

梁井宇
场域建筑工作室
创始人主持建筑师

教育背景
1987年－1991年 天津大学 建筑系 工学学士

工作经历
1991年－1996年
机械部设计研究总院 建筑师
1996年－1999年
加拿大蒙特利尔 Hanganu Architects 及 Campanella Architects 事务所 建筑师
2000年－2002年
加拿大温哥华Electronic Arts Canada 电子艺术家
2003年－2006年
北京中联环建文建筑设计有限公司 总建筑师
2006年至今
北京场域建筑工作室 主持建筑师

主要论著
梁井宇（作者之一）.《平民设计、日用即道》[M]，同济大学出版社, 2016
梁井宇（作者）.《梁井宇——当代中国建筑师系列丛书》[M]，中国建筑工业出版社, 2012
梁井宇（译著）.《庇护所》[M]，清华大学出版社, 2010

设计获奖
2008 – 北京伊比利亚当代艺术中心获WA中国建筑奖优胜奖

代表作品
黔东南茅贡镇旧粮库改造2016（图1）、北京大栅栏文保区保护及发展规划及杨梅竹斜街试点项目2010-2013（图2）、第15届威尼斯建筑双年展中国馆2016（图3）、四川乐至报国寺净苑禅修中心2013-2016（图4）、北京無用生活体验空间2014（图5）

日常经验·个体记忆+居住畅想

三年级建筑设计（6）设计任务书

指导教师：梁井宇
助理教师：青山周平

日常经验与个体记忆

在成为建筑师之前，我们对建筑的喜好来自日常经验和个体记忆。童年玩耍的地方、亲人的 房间、密林、宠物、山川、田原、村舍、街巷，乃至印象深刻的电影、庙会、集市、明信片、 文学作品、音乐，以及弥漫场所周围的气味、光线、声音、材料等等。它们紧紧包围着我们的真实生活，不知不觉中成为头脑中的一部分。我们开始抒发、评论、勾画、模仿、重构头脑中的经验和记忆——一幅画、一幅窗帘、房间布置、街头留影、对旅途中的建筑评头品足……这些汇集而成超越个体的地方经验、集体记忆。

进入建筑学教室后，我们开始接触世界各地的建筑案例、理论、历史、建造技术、材料。现代化的教育理论强大而有说服力，日常经验和个体记忆退缩到大脑隐秘的角落，我们对建筑、空间、材料的判断不再单纯依赖直觉。大师的光环、老师的说教、经典的案例让日常生活中的建筑都显得平淡无奇，我们向往像先锋派大师那样，"发明"崭新的空间，做出"原创"的设计。与此对照的另一面，我们用建筑功能组织、样式、材料、规范、建造技术知识将自己变成专业官僚，怀揣创作欲望，提供专业的解决方案。我们以为这就是理性与情感的结合。

课程目标

因为涉及私密的个人情感，或是觉得谈论这些有违职业伦理（将个人喜好强加在业主的建筑中），建筑师对灵感来源、设计概念的阐述往往回避谈论建筑师的主观影响。相关论述的缺乏导致学生很难通过理论、历史、案例，在课堂进行相关的学习。再加上个体的主观感受又因人而异，因此我们常常感叹，在学校里，"建筑学不可教"。所以许多建筑师的成长需要在师徒制的工作室（建筑或手工艺学徒）中，通过明师的言传身教及其周围的工作生活，耳濡目染，将这种宝贵的灵感财富运用到设计中。这几乎是建筑师成长的必由之路。

本次设计课程就是要尝试在课堂教学中传授，将"不可教"的内容变成"可以学"的主观经验的运用。培养学生树立设计的价值观，学习如何观察自身的日常经验和个体记忆，并进行表述，并将之作为设计灵感运用在居住空间的设计中。

课程描述

本次设计课程分成三个阶段，在第 1-3 周课程中，每位同学被要求描述清楚各自真实日常生活中（而不是旅游偶遇的特殊空间）感受强烈的空间、物件（家具、器皿、服装、玩具、庭院小品等人工物），以及它所蕴含的情感。描述的方式可以是文字、徒手画、照片、影像、声音、或者是以上几种媒介的混合物；之后第 4~6

周课程则是将以上个人化的空间情感体验带入相关的建筑空间，具体要求是对学生特定选择的人士（身份需要详细描述）的居住空间的一项功能（比如睡眠）或两项功能（比如准备食物+蔬菜花园）的设计（不需要完整的居住功能）。最后第 7~8 周，除了标准的设计图面表达之外，学生还需要使用在上段课程中使用的媒介（文字、徒手画、照片、影像、声音、或者是以上几种媒介的混合物）对作品进行最终解释。因此，完整的设计作业可以理解为有三部分内容构成，并在最终汇报时作为整体呈现：（1）起因：意图与设计冲动（即原始的空间情感描述）；（2）过程：设计的内容(居住空间设计)；（3）表达：意图在设计中的再现（即空间的情感是如何在设计中实现的）。

设计要求
本课题以学生选定的具体的一个人或一个家庭为设计服务对象，为他(她)们设计满足其居住的某一、或某几项居住需求的空间设计。设计要求及设计条件如下：
1. 所选择的设计服务对象的身份描述与其居住行为的分析研究；
2. 所选定的居住需求的具体分析、案例比较与研究；
3. 该居住空间及内外家具、庭院、植物、动物(如有)的尺寸的比较研究和最终确定；
4. 1:50或不小于该比例的平面、立面、剖面图纸(含家具、人、动物、植物)；
5. 在课程描述中要求的，运用其他媒介对设计意图的表达；
6. 本设计不设基地要求，但是如果学生认为和课程目标有关，也可以自行选定基地，但是基地不能成为本次设计课程决定性的设计因素。举例来说，一般不应选择自然地貌特征明显的海边、坡地等，或非常规形状、位置的城市内用地构筑设计方案，但是如果学生所需要表达的过去经验中"海""山坡"是其中不可分割的重要空间组成时，方可纳入被动的设计的要素进行考虑，但衡量方案的设计水平还在于建筑本身，即设计人主观可控制的部分。

设计成果表达
1. 日常经验和个体记忆的表达
2. 居住空间对象的研究与分析
3. 居住功能的研究与分析
4. 建筑材料、色彩的表达
5. 空间与尺度的表达
6. 运用综合媒介对空间情感的表达

桌/HOUSE WITH A TABLE

项目选址：无
项目类型（功能）：住宅
建筑面积：171m²
用地面积：242m²

方案设计：刘潇潇
指导教师：梁井宇
完成时间：2014

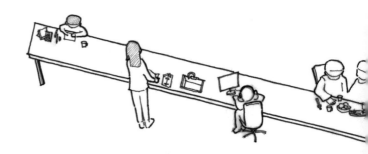

开篇：主要方案图。**本页上图**：关于一张桌子的想象。**本页下图**：平面图。**对页上图**：轴测图。**对页下图**：轴测图轴测图。

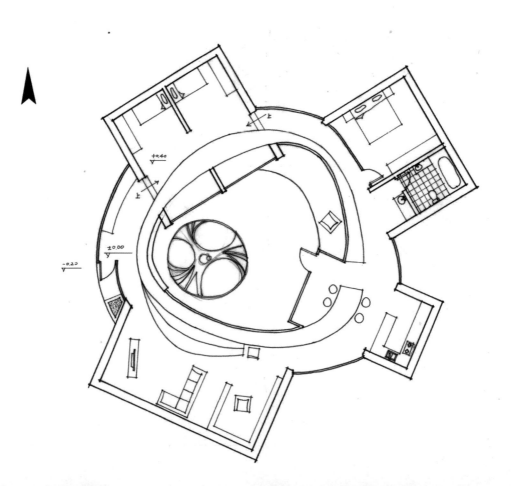

我的设计想法源于对于"饮食"这项活动的观察。在中国传统文化中将"食"与"礼"建立起联系,"吃"在传统的宗族家庭文化中占有举足轻重的地位。然而现代城市生活中,忙碌的工作学习占用了大部分时间,家庭成员能够彼此交流、共度时光的机会往往只剩下晚上的一餐饭。在我的家中,一家人很少有机会能够坐在餐桌边共进晚餐,往往在各自房间中就餐,同时伴随看电视、看书、工作等活动,餐桌的存在形同虚设。这启发了我,能不能通过某种特殊的方法,让家庭成员在保有各自隐私的情况下,增加彼此的交流?餐桌在家庭生活中扮演者怎样的角色?它能否承载更多的家庭使用需求?能否通过对于餐桌及其周边空间的塑造,来定义整个住宅中其他的空间?我的想法是通过一张形状特殊的桌子将传统的封闭的家庭房间切开,形成一个贯穿整个家的环形公共空间。由于环形特殊的方向性,通过向心与离心能够形成介于封闭与开放之间的不同的空间体验,在保持家庭成员各自独立的生活的同时增加声音或视线的交流。设计最终回归到个体的生活记忆,通过将生活场景置入具体的空间中,让空间变得生动。

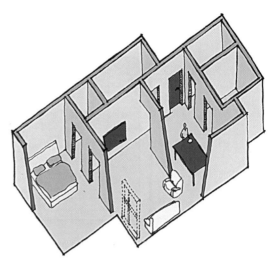

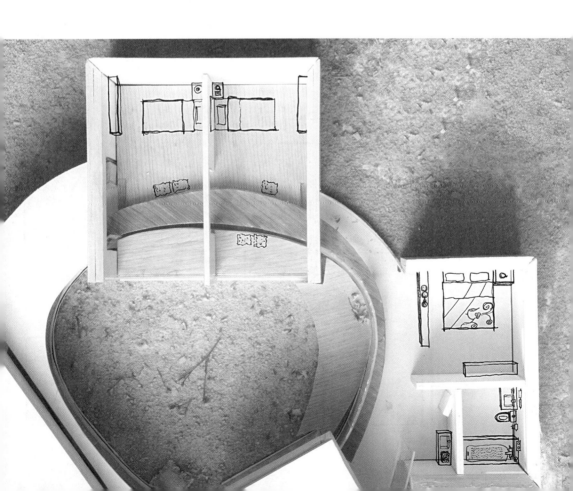

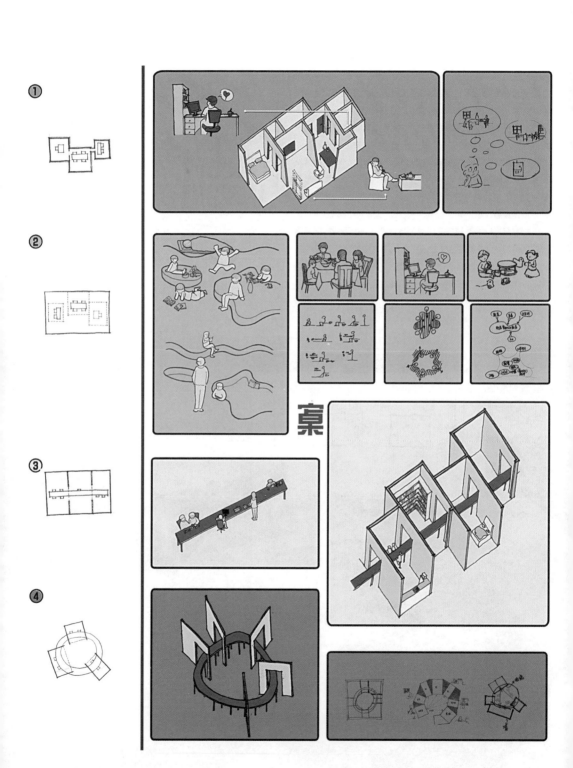

对页图：方案发展过程草图。**本页图**：生活场景与模型拼贴。

My design idea originated from the observation of the activity of the "diet". In Chinese traditional culture, the "food" and "ritual" have established contact, and "eating" occupies a pivotal position in the traditional family culture. However, in modern urban life, the busy work and study take up most of the time, and the family members can communicate with each other and spend time together during the night meal only. In my home, my family rarely have the opportunity to sit together to share the dinner, we often eat in our own rooms, accompanied by watching TV, reading, working and other activities, and the table performs practically no function. This inspires me. Could the family members increase the exchange of each other while maintaining their privacy through a special way? What is the role of the dining table in the family life? Can it carry more functions in family life? Can the dining table and its surrounding space be shaped to define the rest spaces of the entire residence? My idea is to separate the traditional closed family rooms through a special table, forming a circular public space that runs through the whole family. Due to the special circular direction, different spatial experience between the closed and open degree can be formed through center-facing and centrifugation, which can increase the communication through voice or sight while maintaining the independent life of family members. Design eventually returns to the memory of the individual's life, the space becomes vivid through putting the life scenes into the specific space.

对页图：庭院鸟瞰。**本页上图**：房间之间的对视关系。**本页下图**：起居室透视图。

教师点评

从刘潇潇完成的方案很难看出各自对生活观察理解的"费劲儿"。事实上，这是她们花费了课程一半以上时间反复尝试、摸索出来的。刘潇潇在设计开始时，几乎无法描述自己生活中的细节，生活抽象化为简单的时间作息表，而没有对空间具体的感受。"一家人围坐在一起吃饭"的设计意图是通过反复启发和自我检讨，终于发现自己和父母在家吃饭时"都是各自找地方，很少有在一张桌子上一起吃饭"这个与空间有关的问题。找到这个看似简单，但真实有效的设计问题，事实上便是实现了这个课程的教学目的之一。令人有些意外的是像她同样的情况在组里还有不少，普遍对自己"熟悉"的生活观察较为空洞而概念化，缺少真实的鲜活的感受。好在所有同学都极为善于学习和沟通，通过简单的几周课程，都有了显著的进步。

Teacher's comments

It is difficult to make out the "great efforts" taken to observe and understand the life from the scheme completed by Liu Xiaoxiao. In fact, it takes more than half of her course time to obtain the result through repeated attempts and exploration. At the beginning of design, Liu Xiaoxiao could hardly describe the details of the family life. The life abstracts to the simple time schedule, without specific feelings to space. The design idea of "a family sits at a table for meals" comes from repeated enlightenment and self-criticism. She finally finds the issue about space that she and her parents "find a separate place to eat, but hardly sit at the same table". As a matter of fact, finding of this seemingly simple but real and effective design concept helps achieve one of the teaching objectives of this course. Unexpectedly, there are many similar situations in her team. Commonly, the observation to their "familiar" life is relatively inane and conceptual, lacking real and vivacious feelings. Fortunately, all students are extremely good at learning and communication, and get obvious progress through several weeks of simple courses.

方案设计：赵慧娟
指导教师：梁井宇
完成时间：2014

本次设计课题从回忆自己的日常经验开始，经历过对自己过往经历连续几周探寻后，提取出记忆中最为深刻的那段，反复琢磨才领悟到期间的美好都源于淳朴的生活——世代同居，隔代亲情渗透在日常生活的点点滴滴。如何把这种不可见的感受物质化到空间中？如何把人的体验、对生活的点滴理解升华为表达某种共性的设计？

感性的体验经过抽象逻辑的提取，完成从现象到本质的自我问答。最终归结为在世代的家庭结构下共居与独居关系的探寻上。思考了这样一种理想环境：随着世代家庭逐渐被社会发展拆解细分，家庭细胞逐渐缩小，本次设计试图发展出一种新的可能模式，呼唤传统家庭结构的回归，实现多世代家庭中亲情和交流的归位，同时满足各自家庭的隐私需要。最终再现个人独特经历，引发共鸣。

开篇：主要透视图。**本页图**：一层平面图。**对页上图**：居住模式研究。**对页下图**：模型效果图。

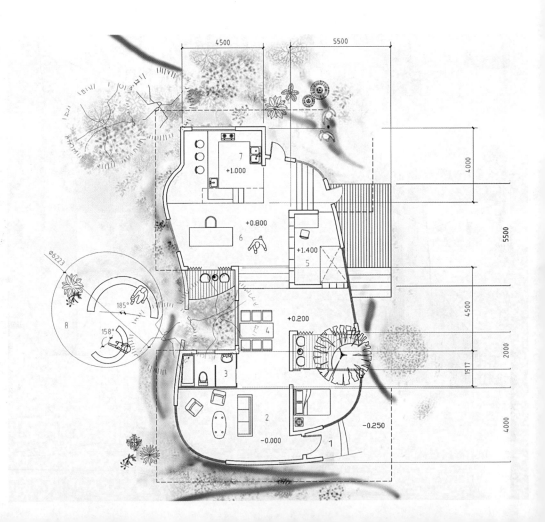

在传统家庭结构下多为三世、四世同堂。居住模式多为传统的"四合院"类型。即如图中表示，不同代人在平面上有各自独立的生活区域满足个人生活、隐私等需求，中心区域则是开放的、多代人一起在此共享的空间（楼房的结构也是如此）。这种模式下三代人之间的交流可能被平面上的"房间"阻隔，分别待在个人的房间里，且平铺的面积很大。

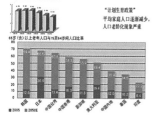

竖直方向分出公共与私密，分出楼板上下不同空间结构——下层完全开放，上层各自独立，各有独立垂直交通。——一个家就是一个微观社区。

考虑到祖父母上楼不便，对他们来说设计成慢慢抬高的设计，楼板标高分段抬高，不会太多不便。

本页上图：模型鸟瞰效果。跨页图：模型庭院效果。对页上图（从上至下）：剖面透视图。二、三层平面图。

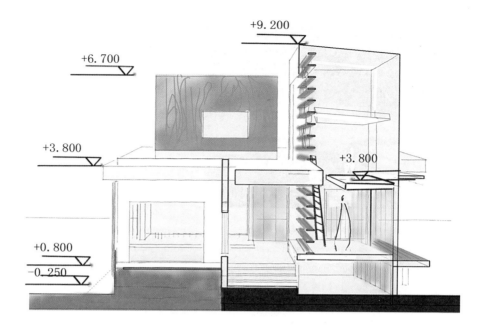

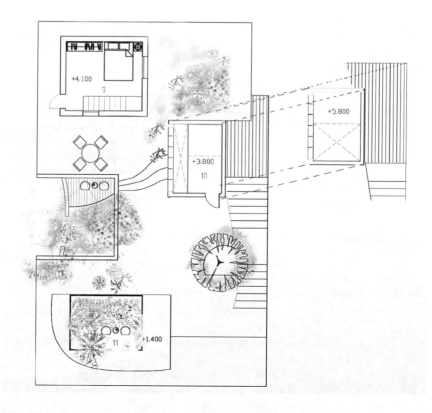

This design topic starts from the memories of our own daily experience, after analysing the past experience for several weeks, we extracted the most profound period from memory and conduct repeated pondering. We realized that the happiness is derived from the simple daily life – different generations living together and the inherited kinship penetrates in daily life. How to turn this invisible feeling into space? How to upgrade people's experience and the understanding of life into the design that expresses some shared values?

Through crystalizing the abstract logic, perceptual experience complete the process of self-criticizing. It finally comes down to explore the relationship between the cohabitation of different generations vs. and independent habitation. Imaging an ideal environment: with multi generations families are decomposed as society develops, the family cells gradually reduced, the design tries to develop a new possible that calls for the return of the traditional family structure, bringing back the affection and exchange in a multi generation family and still satisfy the needs of the privacy of their homes at the same time. Finally produce the unique experience that all individual resonance.

本页上图：原始概念。**本页下图**：功能地图。**对页下图**：主要方案图。

教师点评

赵慧娟在一开始的观察描述中,很快建立起自己对楼梯的兴趣。但不是很明确这种兴趣的来历,究竟是因为楼梯带来空间的转换改变(楼上和楼下空间因为楼梯而产生区隔或间接联系),还是楼梯本身的空间趣味性(楼梯本身的空间形态)。这也是我颇费周折试图让她弄清楚的地方。渐渐地,她将上下楼的行为和农村含有三代人口的大家庭生活关系结合起来,打开了一条研究大家庭内部公共和私密生活的新思路。之后的设计就比较轻松了:一个开敞的首层分别有独立的楼梯到达各自私密的二层小屋,而各自二层的小屋之间通过户外屋顶花园区隔。赵慧娟成功而不自觉地将"微型社区"的形态引入到一个家庭的内部。但略微可惜的是,设计本身的结构还显得比较松散,表达方式也有待进一步提高和完善。

Teacher's comments

After the initial description of observation, Zhao Huijuan soon establishes her interest in stairs. But she cannot define the origin of the interest, that is, whether the spatial transformation and change are caused by stairs (spaces upstairs and downstairs are separated or indirectly connected by the stairs), or the stairs are provided with spatial interestingness (spatial form of the stairs). It's also the point I take much effort to make her to be clear with. Gradually, she combines the behaviors of going upstairs or downstairs with the living relations of the big rural family containing three generations, opening a new line of thought to research public and private life in the big family. The following design process is relatively easy: there are independent stairs to reach respective private rooms in the second floor from an open first floor, and the respective rooms in the second floor are separated by outdoor roof gardens. Success of Zhao Huijuan is the unconsciously introduction the form of "micro-community" into interior of a family. But it is a little bit pity that structure of the design is relatively incompact, and the expression requires further improvement and perfection.

依树居/LIVING BY TREES

项目选址：虚拟街区
项目类型（功能）：住宅
建筑面积：300m²
用地面积：600m²

方案设计：唐紫晔
指导教师：梁井宇
完成时间：2014

依树居是以树为主题出发的住宅设计，源于作者对住所院外树下的生活场景，如吃饭、赏月、歇凉的记忆。树梢的蝉鸣、叶隙的阳光、枝上的落雪、地面的斑驳、风吹的摩挲，树下的感观体验随着季节、时刻、天气变换。树的有机形态，枝叶的清香传递着自然气息。在城市里，拥有一颗树，就好像拥有了自然。一棵树也常承载着几代人情感的延续。树自身的优良特性，比如形状与空间的分支方式、彼此保持合适距离、与自然充分接触与相处等等，是建筑与生活当学习之处。树为居住带来自然气息，人们也总是希望让居住贴近树。树屋凌驾于树之上，尽管与树亲密，但或许对树占用太多；而在一些围廊、天井、四合院中，建筑与树则以适当的距离，保持相互尊重与借用的关系。建筑与树间的关系如何把握，建筑如何尊重树的存在，也需要建筑师仔细思考。因而树与居住的关系成为了设计的主要关注点。

依树居设想住家为一个三代六口人的家庭，在城市街区一片种植有两棵树的场地，以树为主体，建筑为客体，建筑使用树剩余的空间，形成S形平面。建筑高三层，让居住者从不同高度、不同距离观看两棵树。剪力墙由下向上如树般生长，作为结构，同时引导着观树的视线。剪力墙形成的居住空间，也从下而上，由公共（会客、用餐、烹调、读书）向私密（睡眠、沐浴）分支。

开篇：主透视。本页及对页图：概念模型。跨页图：分析——树与建筑。

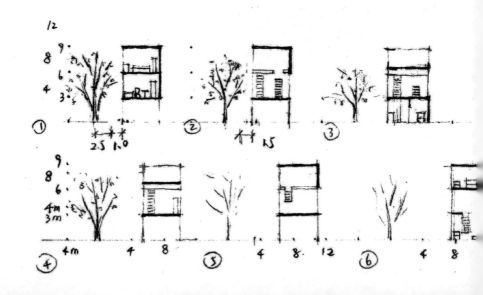

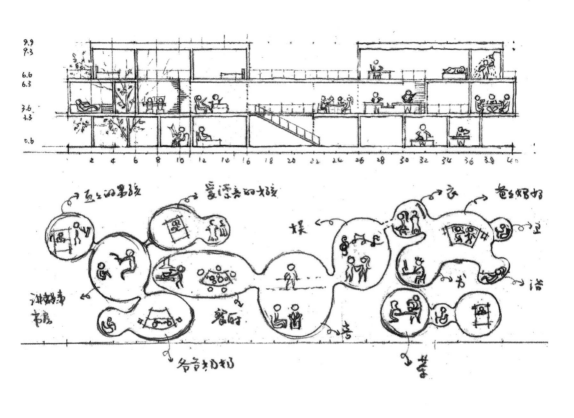
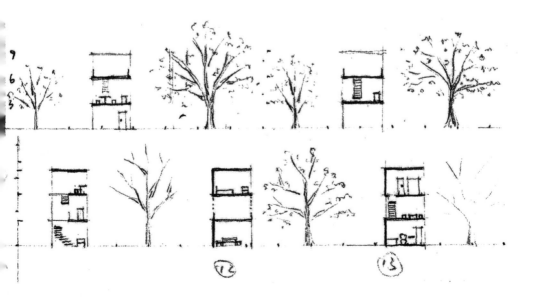

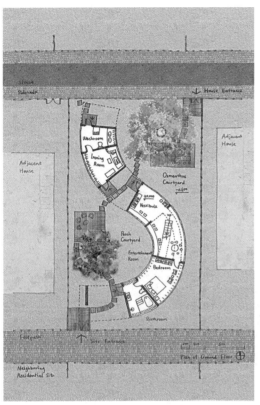
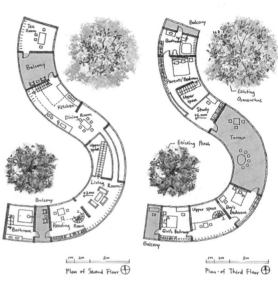

本页上图：平面图。**对页上图**：剖透视。**跨页图**：过程。

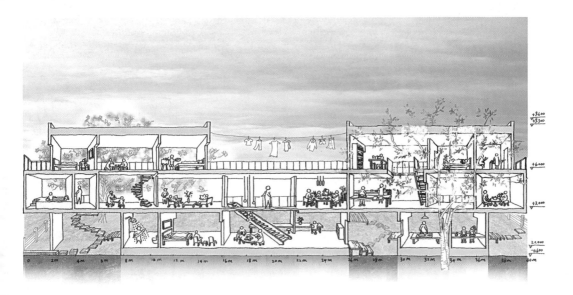

Living by tree is the residential design that takes the tree as the theme, which stems from the scenes of life under the tree outside the shelter, such as the memories of eating, enjoying the full moon and enjoying the cool breeze, cicadas from the tops of trees, sunshine from leaf gap, snow fallen on the branches, mottle on the ground, the wind caress and the view experience under the tree change with season, time of day and weather. The organic form of the tree, the fragrance of branches and leaves pass the natural atmosphere. In the city, with a tree, it seems to have a piece of nature. A tree often carries the continuation of several generations of people's emotions. The outstanding characteristics of the tree, such as the branches of shape and space, maintain a suitable distance to each other, have full contact with nature, and so on. Trees bring natural atmosphere for living, and people always want to live close to the tree. Tree house rises over the tree, although being close to the tree, it sometimes occupies too much on the tree; and in some corridors, patio and courtyard, the building and the tree have appropriate distance, maintaining the relationship of mutual respect and borrowing. The architects should think carefully about how to grasp the relationship between the building and the tree and how to respect the existence of the tree. So the relationship between the tree and the living has become the main focus of the design.

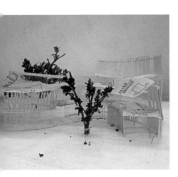
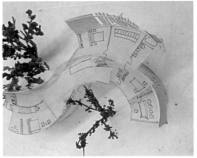

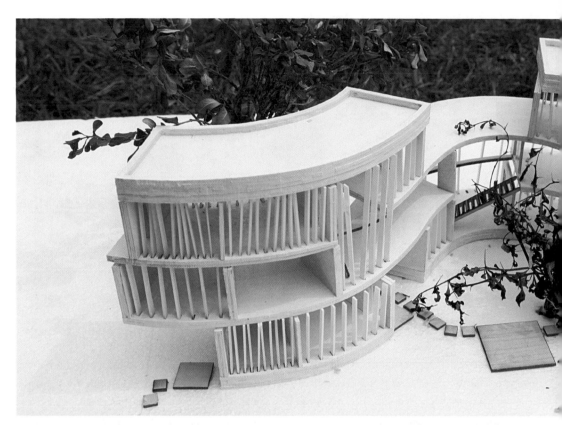

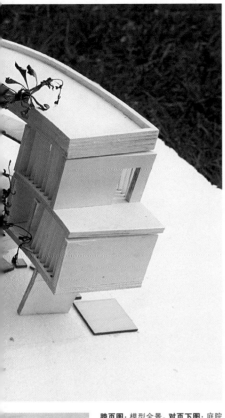

教师点评

唐紫晔代表了另一类别同学,他们不缺少现实生活细节的观察、记忆与再现能力,头脑中充满过往场景与氛围的储备。然而,当需要从这些丰富的生活素材中提取有效的设计源泉时,他们往往不知如何取舍,或误取些与空间无关的内容。从唐紫晔最初提出自己对"院子"的喜爱开始,许多虽然是喜欢院子生活的缘由,但也有与空间无关的因素被提出。找到真正令她感到喜爱的关键性空间元素——"树下空间"却是经过了不少的周折。唐紫晔的设计体现的另一个设计问题也是普遍性的,即如何不照搬现实生活,而是利用对现实生活中美的、有价值的空间的观察,将它转换为设计方案。从最后完成的方案看,将一个原始而简单的"树下空间"作为她的设计原点,将它发展成为一个与树相处的建筑,目的基本达到了,但还是看得出结果非常辛苦而造作。当然,作为建筑师,我们一直都在努力,永无止境的工作不就是要将这些辛苦而造作的设计变得趋近轻松与自然么?

Teacher's comments

Tang Ziye represents another type of students who do not lack the capacities of observing, memorizing and reproducing details of real life. They have the transient scenes and atmospheres stored in their minds. However, they always don't know how to select effective design resources from these rich materials of life, or they may wrongly select some contents irrelevant to space. From the time Tang Ziye put forward her love to the "courtyard", factors irrelevant to space were put forward, although there were many reasons for their love to the courtyard. Finding of the critical spatial element indeed loved by her – "space under the tree" experiences a lot of setbacks. Another design issue reflected by the design of Tang Ziye is universal. That is how to transfer the observed beautiful and valuable space in real life to design scheme, without copying the real life. From the final scheme, she has basically achieved the purpose of developing her design origin of an original but simple "space under the tree" into a building harmonious with the tree. But it can be seen that the result is very hard and intentional. Of course, as architects, we still dedicate our efforts. The endless work is to make these hard and intentional designs become relaxed and natural, isn't it?

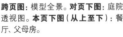

跨页图:模型全景。对页下图:庭院透视图。本页下图(从上至下):餐厅、父母房。

共享书宅/READING —HOME

项目选址：虚拟地段
项目类型（功能）：住宅
建筑面积：300m²
用地面积：1000m²

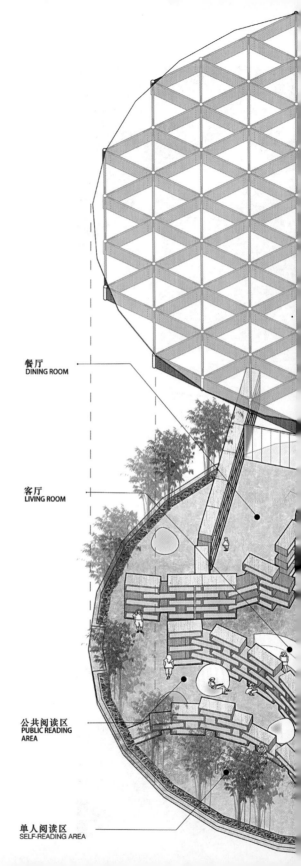

餐厅 DINING ROOM

客厅 LIVING ROOM

公共阅读区 PUBLIC READING AREA

单人阅读区 SELF-READING AREA

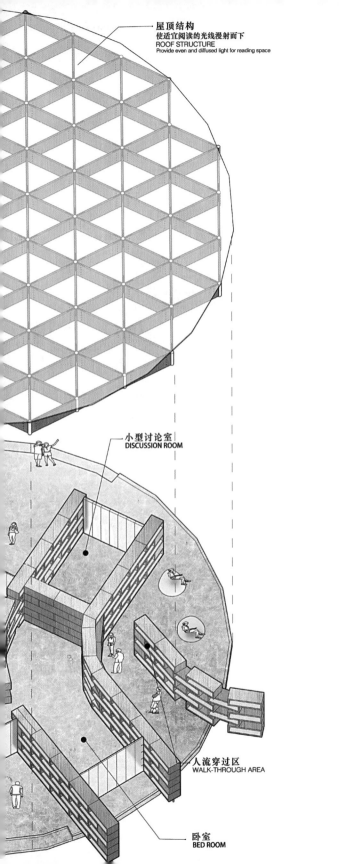

屋顶结构
使适宜阅读的光线漫射而下
ROOF STRUCTURE
Provide even and diffused light for reading space

小型讨论室
DISCUSSION ROOM

人流穿过区
WALK-THROUGH AREA

卧室
BED ROOM

方案设计：黎雪伦
指导教师：梁井宇
完成时间：2014

开篇：主要方案图。**本页右图**：阅读记忆缘起。**本页下图**：功能分区及流线分析。**对页**：方案发展过程草图——设计演化。

缘起于个人的阅读经历，设计试图塑造被书籍所包围的阅读体验。书宅为一位爱书人设计，为他在繁忙的都市生活之中提供阅读的理想空间，并能满足他分享图书的愿望。

蜿蜒的墙面在地上自由地穿梭，划分出或宽敞或逼仄的空间，任意移动的桌椅可以供读书者随时在认为舒适的空间停下，阅读或聊天或只是发呆。

来自不同方向的行人在书架之间穿行，放在书架上的书本身变成空间半透明的分割，透过书架可以感受到对面人的活动，听见外界的声音，天气晴朗时能感受到阳光的照射和空气的吹拂。

书架之间上演着丰富的活动场面，不足1m宽的书架之间可提供一个人单独阅读的封闭与安定空间，1.5m宽的空间供密友促膝讨论或母亲与孩子共同阅读，2.5m宽的空间可以在桌前认真研读，在一颗树附近的空间可供阅读疲惫时的散步，小型讨论室可供读书小组或主人会客使用，向外开敞的空间也可作为二手书贩们的临时卖场。而主人在其中可以阅读，也可以会友。

关于如何将对于读书空间的抽象记忆转化为建筑空间，在方案生成过程之中做过许多种不同的尝试。从最初关于具体的一个空间的想象，到不同的建筑原型的设想，其中包括起伏而连续的地面、独立的一组阅读小屋、迷宫般构成的书架等。在这个过程之中，试图利用建筑语汇清晰而准确地传达阅读空间的精神。最终方案选择了蜿蜒曲折的曲线形书架，以及匀质的圆形屋顶作为建筑语汇，是为了增强整个空间的流动性和统一性，使方案感念被更加纯粹地表达出来，阅读空间体验也更加突出。

如何将住宅功能和公共的阅读功能相结合，是方案面对的一个矛盾点。在最终设计之中，住宅功能被分散成一个一个的功能房间，由书架所围合，中央的客厅可做主人客厅以及公共阅读室双重功能使用。每个功能空间都朝向外部不同方向的景观，并具有一定的私密性。方案提出的解决策略几乎是架空的，地段也非实际地段，在这次设计课程之中，解决实际问题的需要被暂时搁置，而不断捕捉特定的空间体验才是关注的重点。

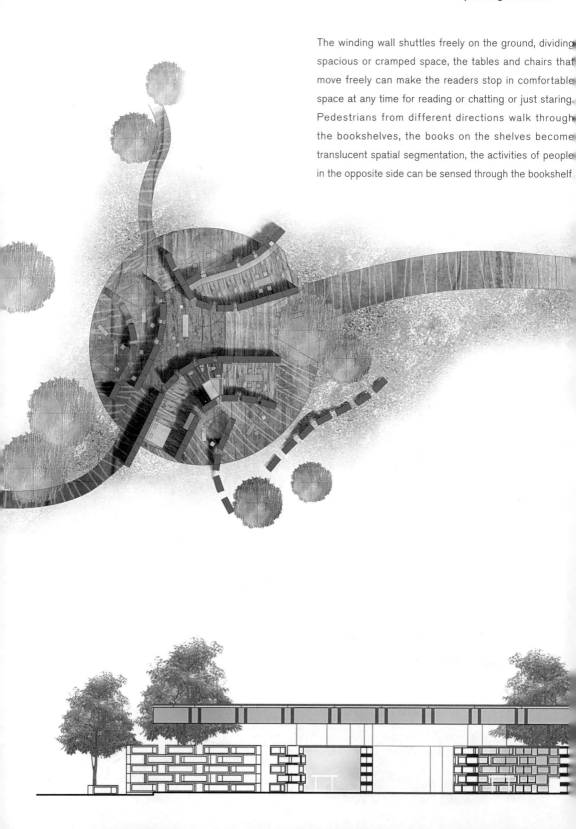

The winding wall shuttles freely on the ground, dividing spacious or cramped space, the tables and chairs that move freely can make the readers stop in comfortable space at any time for reading or chatting or just staring. Pedestrians from different directions walk through the bookshelves, the books on the shelves become translucent spatial segmentation, the activities of people in the opposite side can be sensed through the bookshelf.

and the voice can also be heard, the sun exposure and air blowing can be felt in good weather. The rich activities are staged between the shelves. the closed and stable space for individual reading can be provided between the shelves less than 1 meter wide. 1.5-meter-wide space can be used for discussion among close friends or shared reading between mother and child.2.5-meter-wide space can be used for careful reading in front of the desk. The space near a tree is available for walking after tired reading. The small conference room can be used by the study group or entertaining the visitor by the host. The outward open space can also be used as a temporary store for second-hand book traders that the host can read and meet friends in it.

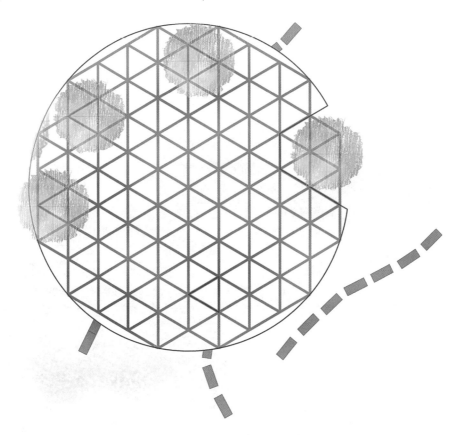

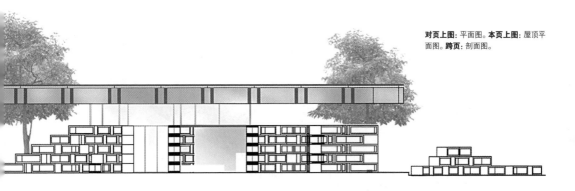

对页上图: 平面图。本页上图: 屋顶平面图。跨页: 剖面图。

本页图：阅读空间透视图。**对页图**：厨房、餐厅空间透视图。

教师点评

这个方案的日常经历来自黎雪伦同学平时爱读书的习惯。这个经历与空间的关系可近可远，不好把握。在最初的个人经历的描述中，她也面临和刘潇潇相似的问题，对自己的个人生活体验理解较为抽象，缺乏空间感受。在对自己一位长辈的书房进行了较为细致的调查和图解后，空间、家具、人的使用行为变得具体而丰满起来。于是她很自然地把图书、阅读、书架、桌椅家具、户外环境等等几个方面联系了起来，运用较为娴熟的形式操作手段，顺畅地给我们带来了一系列完整的空间体验。从室内到室外空间的融合，最终实现了一个可以让人充分享受的阅读环境。这是一个理想的从日常经历到设计空间的转换案例。但是这种近乎完美的答案与现实生活的不完美的反差令人有些不安，这容易滋生建筑师介入现实的一种唯美错觉。显然，现实生活中是不容易有这样轻而易举的设计答案的。如何正确地理解现实，并有效地提出针对现实的解决方案，是黎雪伦下一步需要面对的设计挑战。

Teacher's comments

Daily experience of this scheme is from Li Xuelun's love of reading books. Relation of this experience and space could be close or far, which is not easy to grasp. In the description of initial personal experience, she also faced the similar problems as Liu Xiaoxiao, who had a relatively abstract understanding to her personal life experience and lacked the spatial feeling. After relatively detailed investigation and illustration for the study of an elder, the space, furniture and usage behavior of people, the design become concrete and well-rounded in her course description. Therefore, in the later design, she spontaneously connects several aspects such as books, reading, bookrack, tables and chairs and outdoor environment, and swimmingly together. She brings us a series of complete spatial experience by taking advantage of the relatively skillful formal method of operation. The integration of indoor and outdoor spaces helps finally achieve a reading environment that may be fully enjoyed by people. This is an ideal case of transferring from daily

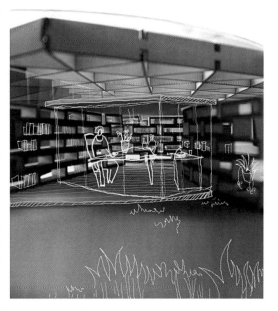

experience to design space. But the almost perfect answer and the imperfect real life make us uneasy, because it's easy to generate a kind of beautiful illusion that the architect pictures in the reality. Obviously, it is not easy to generate such easy design answer in real life. Facing the design challenges in the future, Li Xuelun shall correctly understand the reality and effectively, then put forward the solutions for the reality.

卧居/SLEEPING-HOME

项目选址：无特定地点
项目类型（功能）：住宅
建筑面积：280m²
用地面积：500m²

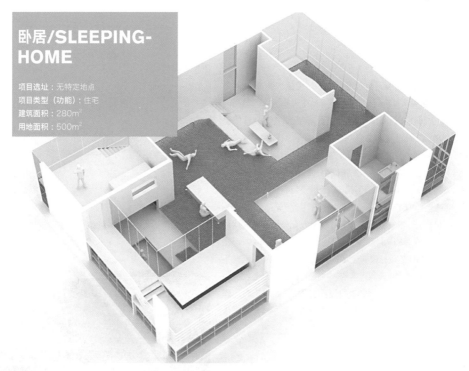

方案设计：戴锐
指导教师：梁井宇
完成时间：2014

该方案从日常生活中对睡眠空间的体验出发，以睡眠姿势、视线、材质等角度着手，设想了残疾人住宅设计的一种可能。在住宅中，利用藤编的网状平面营造出一层柔性活动空间，以450mm为单位进行空间错层，既方便家人照顾，又在一定程度上保护了残疾人生活的私密性。同时在书房、浴室、餐厅等处加入了针对残疾人的细节设计。

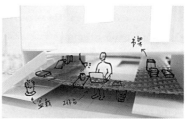

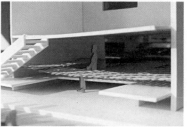

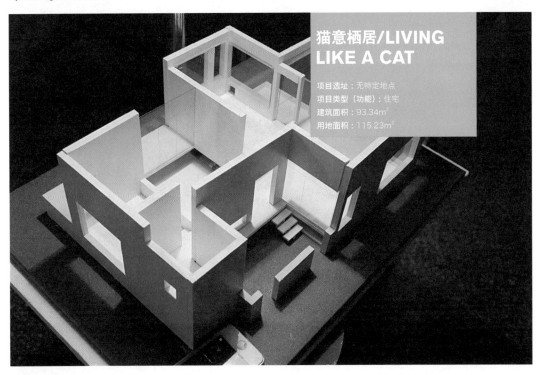

猫意栖居/LIVING LIKE A CAT

项目选址：无特定地点
项目类型（功能）：住宅
建筑面积：93.34m²
用地面积：115.23m²

方案设计：刘畅
指导教师：梁井宇
完成时间：2014

猫意栖居缘起于在老宅与家猫相伴的几千个日夜，猫与人隔离在室内外的两个空间内，凭借各自的声音与肢体语言努力交流，偶尔亲密接触，却引发各种哭笑不得的画面。日后回想来，猫与人都迫切地想与对方共享温馨的生活，但仅关注人的空间设计阻止了愿望的实现，同一屋檐下的生活反而变得状况百出。

因此方案从猫与人的共存视角出发，为彼此预留私密尺度与空间的同时打开隔阂，人尝试着以猫的视角看待空间，以猫的方式享受生活，而猫则拥有了自己的"楼层"与动线，二者得以在小小屋檐下诗意栖居。

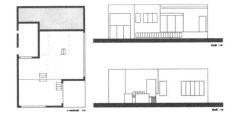

回归自然·消融边界 / BOUNDARY ABLATION

项目选址：中国南方（四川东南部）某城郊
项目类型：住宅
建筑面积：300m²

方案设计：李嘉雨
指导教师：梁井宇
完成时间：2014

这是一个实验性质的方案，设计伊始，指导老师梁井宇老师要求我们从"日常体验·个人记忆和居住理想"的方面来考虑设计概念和思路。于是我从童年的个体记忆里寻找到了幼时在四川东南部乡村生活的一段体验，并通过时间和大脑的发酵想象与保留，形成了现在对那时生活的一种印象。通过这样的印象，我绘制了一些手绘图来表达我的设计意向，并通过查找实际的建造资料，将这些意向提炼为"把居住空间的边界打破"。基于这个概念，我设计了一栋住宅，它在中国南方的气候里需要保持凉爽、通风、适度的明亮、视线良好等等条件，更重要的是，它能依据房间的不同功能，将其边界——最基本的是块体的6个面——以合适的方式进行不同程度的消融。这就需要结合当地的材料、工艺等等条件对不同的边界进行结构、构造、工艺的种种设计，以达到人类居住功能和消融与自然边界的最佳平衡。

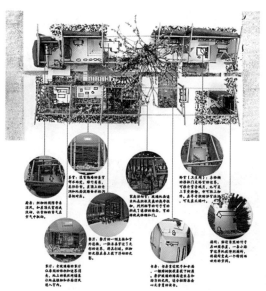

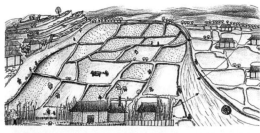

小型悬挂住宅/ SUSPENDED HOUSE

项目选址：悬挂于立交桥下
建筑面积：30m²
用地面积：30m²

方案设计：刘少文
指导教师：梁井宇
完成时间：2014

解决城市中众多年轻人生活乏味的问题。
通过悬挂住宅的概念，利用城市的边角空间提供有趣的生活场所。

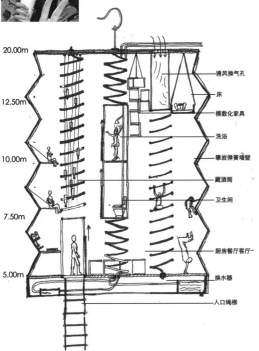

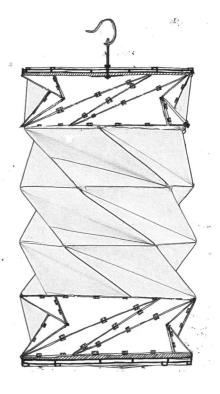

寻回童年的院子/
COURTYARD FOR CHILDHOOD

项目选址：北京旧城胡同
项目类型：居住、改造
建筑面积：309.5m²
基地面积：370m²

方案设计：薛昊天
指导教师：梁井宇
完成时间：2014

从对童年宅院居住模式的再思考出发，对比了宅院的聚合性和城市单元房的隔绝性，提取出宅院藉由选择性地开放与遮掩创造舒适的半私密半公共空间的特点。在设计中，通过墙元素的围合与错位以及桥元素的连接形成了一个个具有方向性的空间。将这一抽象模式落实到北京旧城胡同内，得到了最终的四合院改造方案。

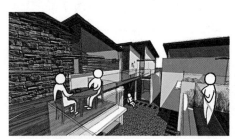

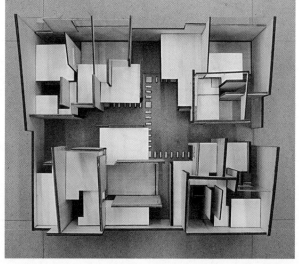

里外/INTERIOR & EXTERIOR

项目选址：无特定地点
项目类型（功能）：住宅
建筑面积：177m²
用地面积：240m²

方案设计：闫博
指导教师：梁井宇
完成时间：2014

儿时在旧仓库"探险"，仓库内丰富空间，不同材质的触感给我留下了深刻的印象。以仓库中空间原型、材质对比出发。用体块错动、特殊材质、丰富光影等建筑的手法创造出相似的空间感受，实现对空间自由灵活的追求。

一层平面图　1:100

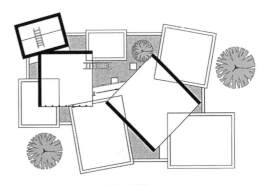

二层平面图　1:100

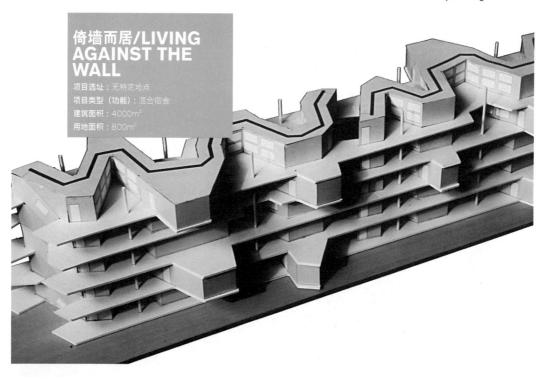

倚墙而居/LIVING AGAINST THE WALL

项目选址：无特定地点
项目类型（功能）：混合宿舍
建筑面积：4000m²
用地面积：800m²

方案设计：袁雪峰
指导教师：梁井宇
完成时间：2014

墙作为一种原型，往往让人联想到分隔、围合、保护、私密等等，但是恰当的运用却能起到增进交流的正面效果，本设计通过"混合宿舍"这个高密度建筑类型，重新思考私密与公共的关系，创造出不同程度的半私密空间，从而丰富居住者的互动与交流。

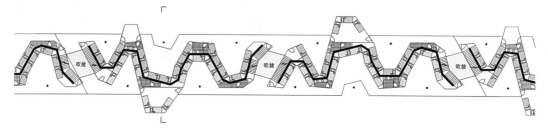

美食·家/HOUSE FROM FOOD

项目选址：无特定地点
项目类型（功能）：住宅
建筑面积：298m²
用地面积：640m²

方案设计：周翘楚
指导教师：梁井宇
完成时间：2014

方案以光影画艺术家为虚拟甲方，以光影画艺术家的工作方式与作品的表现形式为设计要点进行设计。建筑设计的重点在于，如何在建筑外边界被严格控制的情况下，找到合理的空间布局与空间互动，使得建筑充满更多的可能。

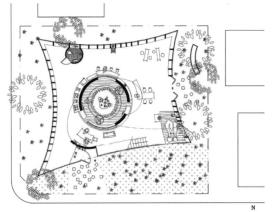

一层平面图 1：50

"光的空间

董功

直向建筑事务所　创始合伙人

建筑设计

董功
直向建筑事务所
创始合伙人

教育背景
1994年
清华大学建筑学院 建筑学学士
1999年
清华大学建筑学院 建筑学硕士
2000年
德国慕尼黑理工大学 交换学生
2001年
美国伊利诺大学建筑学院 建筑学硕士

工作经历
2001年 – 2004年
Solomon Cordwell Buenz & Associates, Inc.
2004年 – 2005年
Richard Meier & Partners, 理查德迈耶设计事务所
2005年 – 2007年
Steven Holl Architects, 斯蒂文霍尔建筑设计事务所
2008年至今
直向建筑设计事务所 创始合伙人

主要论文和著述
《重庆桃源居社区中心》[J]. 建筑细部, 2016.06, P415-418
《Library in Beidaihe》[J]. Detail, 2016.04, P276-281
《Holy Coast- Seashore Chapel》[J]. Architectural Review, 2016.04, P33-43
《三联海边图书馆 直向建筑》[J]. A+U, 2016.03, P40-47
《Chongqing Taoyuanju Community Center》[J]. C3, 2016.02, P88-105
《三联海边图书馆》[M].《中国设计年鉴 Contemporary Design in China 2015/ 2016》, 广西师范大学出版社, 2015, P152-159
《Can Design week save the Hutongs?; Seashore Library》[J], Architectural Review, 2015.11, P23-25, 56-65
《海与光: 三联海边图书馆的两面》[J]. WA世界建筑, 2015.09, P90-99
《Seashore Library》[J]. Baumeister, 2015.09, P20-30
《鲅鱼圈万科品牌展示中心》[J]. Area, 2013.12, P212-217
《CR Land Guanganmen, Green Technology Showroom》[M].《Ecological Urbanism, Harvard Graduate School of Design》, Lars Muller Publishers, 2012, P254-255
《CR Land Guanganmen Green Technology Showroom, Momentary City-CR Land Hefei Dongdajie Sales Pavilion》[M].《Collection of Asian Architecture, Switzerland》, Braun Publication, 2010, P50-51, 56-59
《介入自然-昆山悦丰岛有机农场采摘亭设计》[J]. WA世界建筑, 2012.10, P45-54
《瞬间城市》[J]. Domus国际中文版, 2010.04, P42-45
《CR Land Guanganmen Green Technology Showroom 直向建筑》[J]. A+U, 2009.04, P80-85

设计获奖
2016 – Iconic Awards 公共建筑类 最佳奖
2016 – 意大利Archmarathon最高奖
2015 – Blueprint Awards最佳公共建筑（私人出资）类别特别推荐奖
2015 – A&D Trophy Awards机构/公共类别最佳建筑奖
2015 – 金点概念设计奖年度最佳设计奖（空间设计类）
2014 – 美国《建筑实录》杂志评选的2014全球10大建筑设计先锋
2013 – 全国优秀工程勘察设计行业奖一等奖
2012 – WA中国建筑佳作奖
2010 – 中国建筑传媒奖（最佳建筑奖入围，最佳青年建筑师入围）
2010 – 威尼斯双年展CA'ASI中国新锐建筑创作展，作品征集大赛一等奖

代表作品
三联海边图书馆（图1、2）、海边教堂（图3、4）、重庆桃源居社区中心（图5、6）、苏州非物质文化遗产博物馆（图7、8）、木木美术馆入口改造（图9）、昆山有机农场游客互动中心（图10）、鲅鱼圈万科品牌展示中心（图11）、张家窝镇小学（图12）

"光的空间" 建筑设计

三年级建筑设计(6)设计任务书

指导教师：董功
助理教师：梁琛

光与空间(Space of Light)

人们通常把建筑理解为实体形状和事物的建构，但在历史上和当代很多的优秀建筑范例，显示出对空间氛围的营造，对于无形的，非物质的事物的同等重视。通过难以捉摸的光影的现象，微妙的色调层次，使建筑超越本身的物理极限，在可使用的同时更富有情感。

光在建筑中的核心作用是利用这个力量去塑造人们对于空间的感知，营造与表达生活中精神维度的感受和情绪。课程的重点会致力于光在冥想空间设计中的应用和表现，并探索如下重点问题：光/空间/氛围，光/静谧/情绪/沉思/冥想，明/暗，光/时间/四季，光与影(舞蹈般)的编排。同时也会关注涉及光线条件，场地环境，朝向，气候条件，地方风气，功能需求，建构和建造方式。图纸和剖面模型将在学习中使用，但也要以展示空间与光的关系为主。本设计课程关注于用自然光和建筑空间的关系。在将此付之于具体的功能与场地，并解决建筑设计中基本的空间和结构原则的同时，本课程更注重于建筑中的人对于现象的感知。课程中会有分组研究和独立设计。

课程描述

课程初始将分组进行一些重要案例的分析，研究光在现代建筑中的运用，由此对本类建筑有一个基本的了解。课程最终会将之前的积累运用到设计方案当中，一个在喧嚣的城市或静谧的校园中的冥想空间设计，并以光的呈现为主题。建筑的尺度和场地会被控制，意境，着重于安静和冥想的空间，光与建筑材料和结构的相互作用；架构作为一个有着现象与事件的电影般的载体；光与影舞蹈般地表达空间的秩序和界限；光影动态与色彩的旋律根据太阳轨迹，天气和季节而变换。

冥想空间方案设计将主要由一个大比例的剖面模型来展现，以此将光影以有形的方式表达。学生将以摄影或摄像等方式对模型中的光影做记录。这些将会是评图时的主要表现方式。图纸将被用于以图形方式描绘光影空间，并通过连续的影像，来探讨光和时间的变化以及在空间中的展开。可用电脑表现，但以手工模型和对光影现象的感知为主。

课程目标

1. 案例调研：光的建筑/剖面模型，光影实验为了更好的理解形体中对光的塑造，在课程的第一周将致力于对建成/未建成的光线运用出色的现代建筑案例的熟悉和实验。一方面是为了增加学生对建筑中光的运用的理解，另一方面是为了从分析模型中学习它们如何调节光线，或提出比案例想要得到的更有效的效果，(甚至建议其他并没有出现在建筑中的可能性)。每组学生(两人)将会选择一个光为设计媒介的案例(清单如下)进行分析。主要为探索在体量和开窗中自然光如何被捕捉，引导和操控。在选择的案例中提取一个房间，一个空间片段，或一组空间制作剖面模型，来测试光在此模型中的运用或产生的效果，此模型在不同时间和空间的条件下的光，而不是模型的形态本身。除了模型和分析以外，还需要案例的平面，剖面，概念草图，照片等基本信息。

2.光的建筑：场地分析，方案设计，细部设计，设计表现介绍：场地分析，功能分析，场地模型。

方案设计
第4周的周四中期评图，细部设计完成方案设计阶段后，会在设计中选择一处细部进行深入的设计，此细部可为一组空间，一个房间，一面墙或一扇窗。但必须注重在建构，材料，光影的塑造。比例可不同，可在1:20左右。此大比例细部模型需为最终方案设计的表达重点。

设计表现
此周会致力于完成设计表现：最终模型，图纸，效果图，模型照片等由于在评图时的光线情况不能确定，所以手工模型重要的光影效果需要有照片记录。
第8周的周四最终评图

冥想空间建筑设计要求
(1)建筑选址(学生在两个基地中选择其一)：
a.城市：在北京海淀区五道口区域内，选择一处较为喧嚣的场地作为建筑基地，旨在讨论如何在复杂的城市条件下，创造出城市中以光作为设计核心媒介的静谧的冥想空间；
b.自然：在清华大学校园内，选择一处较为安静、自然的场地作为建筑基地，旨在讨论如何在纯粹的自然的条件下，创造出自然中以光作为设计核心媒介的静谧的冥想空间。
(2)建筑面积：400m^2 (上下可浮动10%)。
(3)建筑内应主要设有冥想空间(250m^2)，并附设门厅(60m^2)、办公室(30m^2)、卫生间(20m^2)、储藏间(10 m^2)等基本功能用房，交通面积30m^{2}左右。
(4)冥想空间可以根据学生的理解自行定义与延伸，在满足个体冥想、静思的基础上，亦可举办如沙龙、诗歌朗诵会、艺术展览、礼拜等社区、校园、文化、宗教活动，可参考禅寺、教堂、咖啡厅、画廊等。

设计成果表达
(1) 案例分析及研究模型；(2) 场地模型；(3) 方案设计研究及模型；(4) 细部设计研究及模型；(5) 最终成果表现。

模糊/VAGUE

项目选址：清华大学近春园工字厅西南侧树林
功能定位：冥想空间
建筑面积：85m²
用地面积：200m²

方案设计：王杏妮
指导教师：董功
完成时间：2014

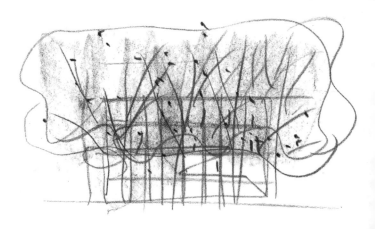

这是一个带有实验性的设计，我企图重新定义自然与人之间互动的边界，希望可以通过一种"模糊的介质"来让人与自然共生。在这个设计里，建筑不再强硬地分割室内外，而成为了某种介质，不仅组织人的活动，也容纳自然的生长，并从精神和物质层面模糊场景内的一切边界。人的行为忽隐忽现，树枝摇曳，真相与影子难以辨别，光作为驱使这一切发生的动因，又与其他元素积极互动。自然与人工交织在一起，创造了一个惬意如诗般的生活场所。

场地是校园中的一片校友纪念林，这里的一草一木都充满着回忆与憧憬。所以，为保护场地的四棵老树，先将一个秩序的纤细框架引入场地，以此让所谓的模糊界面（玻璃钢格栅）漂浮起来。再通过流线串联起功能盒，包括：一人冥想空间、多人休闲空间、休息空间、办公室、卫生间等。光顾这里的人，可能是为了寻得一片安静，也许是为了回忆曾经的故事。 总之，在这里，阳光穿过树枝，透过一层一层的格栅，最终到达了人们的眼睛，一切清晰却又模糊，一切真实却又虚幻，正如回忆般忽的闪现在脑海中，可你却又抓不住这一瞬间的灿烂。

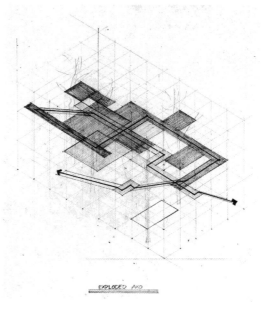

对页上图：概念草图。**对页下图**：室内主透视。**本页上图**：流线轴测图。**本页下图**：总平面图。

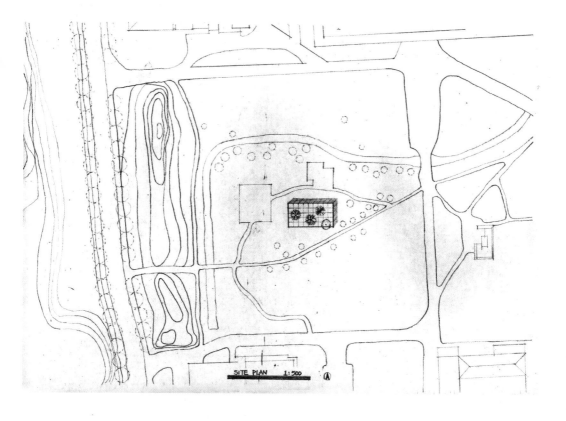

本页上图：模型照片。**本页中图**：隔扇布置轴测图。**本页下图**：室内透视图。**对页上图**：室内透视图。**对页下图**：材料详图。

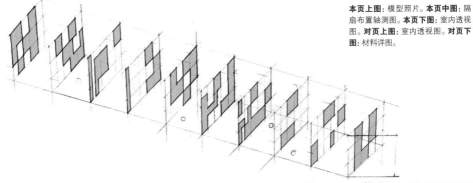

This is an experimental design. I attempt to redefine the boundaries between natural and human interaction, hoping to realize symbiosis between the man and nature through a kind of "fuzzy medium". In this design, the building no longer rigidly divides indoor and outdoor areas, but becomes certain medium, which not only organizes the human activities, but also accommodates natural growth and blurs up all boundaries within the scene from the spiritual and material levels. The human behavior flickers, the branches sway and shadow are difficult to discern, light, as the motivation that drives everything, interacts actively with other elements. Natural and artificial intertwins creating a comfortable living place.

The site is an alumni memorial forest in the campus, where every tree and bushes are full of memories and longing. So, for the protection of four old trees in the site, we first introduce a slender frame into the site, making the so-called fuzzy interface (glass steel grille) float up. Then the function boxes are connected through the circulation, including: a meditation space for one person, leisure space for several persons, rest room, office, bathroom, etc. people who visit here may want to find a quiet place or recall the past memories. Anyway, here, the sun passes through the trees, and eventually reaches the people's eyes through layers of grid, everything is clear and fuzzy, all is real and unreal, like memories suddenly flash back in the mind, but you can't grasp the brilliant moment.

| 玻璃钢格栅

格栅作为一种特殊的半透明材料，用真实的构件赋予了界面虚幻的模糊效果。同时它具有一定强度，可以作为受力结构出现在建筑中。

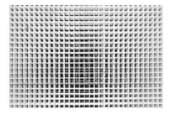

| 高反玻璃

玻璃围合了建筑的透明界面，并在日光照射下反射周边树木、框架，让室内消隐，夜晚却又可以有光亮透出，照亮周边的构件、植物。

| 方钢

由白色金属组成的杆件结构架赋予了场所特定的秩序，并让建筑元素的"漂浮"变成可能；高强度的方钢作为梁柱，让建筑更加轻盈。

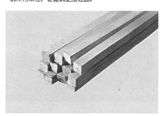

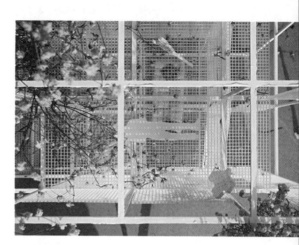

对页图：室外夜景透视图。**本页图**：俯视透视图。

教师点评

该设计颠覆了人们对于光的表现的惯常理解。设计者通过经营场地中的高大树木、细密的构架、以及由两者产生的阴影，让建筑的感受在一天中太阳的轨迹下变幻。设计者起初对于光及由其产生的模糊性的兴趣，并希望通过设计来揭示它。设计以一种最轻的姿态和方式，介入场地，与大树融为一体。人可以从场地的任何一个方向走进这个建筑，并体验由它创造的从未有的场所体验，并在垂直方向上探讨了自由空间及其可能性。

Teacher's comments

This design overturns common understanding of people to the performance of light. Through the management for tall trees and fine structures as well as shadows generated by the both, the designer makes the feelings to the building change with the track of the sun in the whole day. At the beginning, the designer is interested in the light and the generated fuzziness, and hopes to reveal it through design. In the lightest posture and mode, the design enters the site and is integrated with the trees. A person can walk into this building from any directions of this place, feel the unprecedented site experience created hereof, and discuss free space and the possibility in vertical direction. (Written by Liang Chen)

方案设计：张成章
指导教师：董功
完成时间：2014

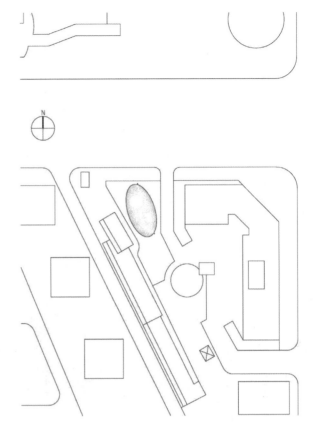

开篇：光的大场景。**本页上图**：总平面图。**对页上图**：平面图。**对页下图**：外景。

Episode I 模糊的瞬间——光·喧嚣

城市生活让越来越多的人带上了眼镜，也让越来越多的人有机会感受到模糊的瞬间。当温热的水汽泛上眼镜，眼前的物体不再清晰，被笼罩在一种朦胧的气氛中。当确定的轮廓逐渐消失，内心也逐渐开始变得澄澈。眼前的世界是否清晰微妙地与时间联系在了一起，当世界变得模糊，时间也似乎放缓了脚步。模糊的瞬间，似乎感受到了上帝的存在。模糊的界面，这是否就是亭子在城市中存在的答案呢？

Episode II 密柱与平板——距离·介质

模糊是一个过程
如果仅仅是界面，即使是模糊的，也无异于一个人工隔离出来的被动的所谓"净土"。我们所设想的是一种介质，随着脚步的深入，与喧嚣的距离不断增加，陷入模糊的程度也不断加深，直到完全蒸发在一个轻盈而静默的屋顶之下。
这种介质并非完全匀质，像烟雾一样，而是与身体相关。在空间当中，密集地分布着细白的柱子，它们的疏密暗示了人在其中行走的流向，最密处刚好比人侧身的宽度略小。
空间与天空的边界是一张平直的薄板。随着距离的变化，城市的模糊程度与空间的整体亮度在微妙地变化着，时现时隐的画面切换与时近时远的焦距终于在抵达空间最深处时不再跳动，这里有一层圆形的玻璃，人体终于在风中寻得了庇护。

Episode III 孔洞与十字——时间·瞬间

在均匀的介质中，光依然可以跳跃
细柱并非被平板直接截断，而是戴上了一个十字梁，嵌在一个圆圆的孔洞里面，光线从平板的上面经过孔洞与十字的诠释而倾泻到柱子上，随着时间的流逝，光斑从柱头流到柱脚，接着又来到白色的地板上。
光斑真的在动，只要你真的在盯着它们看，就像晴朗的秋夜盯着满天的繁星一样。星星在眨眼，光斑也在跳动，当你真的看到了在跳动的光斑，就真的看到了时间，真的忘却了喧嚣，真的找到了安静。

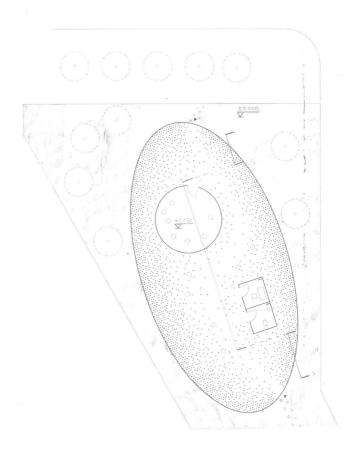

Episode I Vague Moment -- Light · Noise

Urban life helps more and more people wear glasses, but also causes them to feel vague moment. As the glasses are covered by warm steam, the objects in front of us are not clear any more, and we are exposed in a hazy atmosphere. As the definite contour gradually disappears, our hearts also becomes pure increasingly. The world in our eyes is clearly and subtly connected with time, isn't it? It seems that time slows down when the world gets vague. At the vague moment, we seem to feel the existence of God. Is the vague boundary the answer to the existence of pavilions in cities?

Episode II Dense Columns & Flat Plates -- Distance · Medium

Vagueness is a process.

If there is the boundary only, even though it is vague, it will be nothing but so-called "pure land" isolated artificially. What we would like to design is a kind of medium, in which as we enter step by step, the degree of vagueness constantly deepens with the continuous increase of distance from noise till we find that we are absolutely exposed under a light and quiet roof.

This medium is not completely homogeneous just like the haze, but related to the human body. In the space, slim white columns are arranged densely, and their density implies the direction of pedestrian flow. The distance between columns in the densest part is just a bit less than the width of a person on the side.

The space is separated from the sky only with a piece of smooth and level sheet. The degree of vagueness and the overall brightness in the city delicately varies with the distance. Eventually, the flickering picture switch and near-or-far focal distance come to rest at the deepest

space where a layer of round glass gives a shelter for human bodies in the wind.

Episode III Hole & Cross -- Time · Moment
Light can still dance in a homogeneous medium.
A slim column is not directly cut off by a flat plate, but embedded in a round hole with a cross beam, so that light over the flat plate pours on the column through holes after shaped by the cross. As time goes by, light spots flows from top to bottom of the column, and then arrives at the white floor.
You'll find the light spots are really moving as long as you gaze at them just like you gaze up at the stars in a clear night sky in autumn. Light spots are dancing just like the twinkling stars. If you really catch the dancing light spots, you'll really see the time, forget the noise and find the quiet.

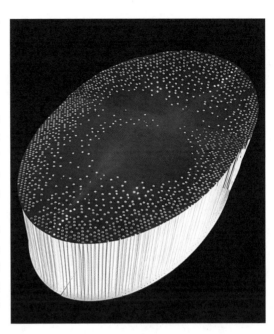

对页图(顺时针)：光的小场景——星星点点。光的小场景——最深。光的小场景——万箭穿心。**本页上图**：晚上的模型。**本页下图(从上至下)**：立面图。剖面图。

教师点评

张成章这个方案的巧思在于结构。密布的柱子分散了每单个柱子的荷载,从而将其截面尺寸削减到尽量小。于是,结构构件开始转向了空间的介质,也是光的介质。柱子之间的间距提示着人在空间中或行走,或停留的行为的可能性,这个设计动作又进一步赋予结构分布某种"身体性"。

从外到内,光环境的由明变暗,制造出空间和现实城市的心理距离。而在这处幽静,甚至是略带幽暗的场所中,阳光会在某个时段穿过柱子与屋顶之间精心预留的孔隙,在空间中洒下密集的,缓慢游移的光斑。这一刹那的景象,犹似把人带入到一百多年前印象派大师修拉的那张点彩传世之作"大碗岛周日的下午"的氛围当中。

Teacher's comments

Zhang Chengzhang' scheme lies in the ingenuity of structure. The close-set columns scatter the load of each single column, minimizing the sectional dimension. Therefore, structural element begins to transfer to medium of space, and also medium of light. Spacing between the columns prompts the possibility of people's walking or staying in the space. This design action further endows the structure distribution with certain "corporality".

From outside to inside, luminous environment becomes increasingly dark, thus forming psychological distance between the space and the real city. In the peaceful and even a little gloomy place, sunshine will pass through the elaborately reserved pore between the columns and roof in a certain period of time, and sheds the dense and slowly moving light spot in the space. The split-second sight seems to bring us into the atmosphere of stippling masterpiece of impressionist master Seurat of more than one hundred years ago – "A Sunday Afternoon on the Island of La Grande Jatte".

跨页图:光的小场景——偶一抬头。

方案设计：秦正煜
指导教师：董功
完成时间：2014

开篇：主要方案图。**本页图（从上至下）**：概念图。地下平面图。剖面图。主要剖面图。**对页上图**：屋顶平面图。**对页下图**：总平面图。

EARTH WOMB是借助围和、下沉的建筑空间和自然光的变化干涉,共同塑造的冥想空间。在设计中,我将人的体验和周围的场所视作冥想的组成部分,而不是把它简单地定义为一个装置。方案的地段位于校园内的草地,子宫状的冥想空间主体沉入地下,给冥想者呵护感,与此同时感受从地面上渗入地下的遥远的光。随着时间的变化,地下空间的光影不断改变、不可捉摸。除此之外,冥想空间顶部的采光筒和它与地面之间的缝隙,能让地面上人的活动通过变幻的光影投射到地下。所以,冥想活动被自然光和地面上人的行为同时激发。

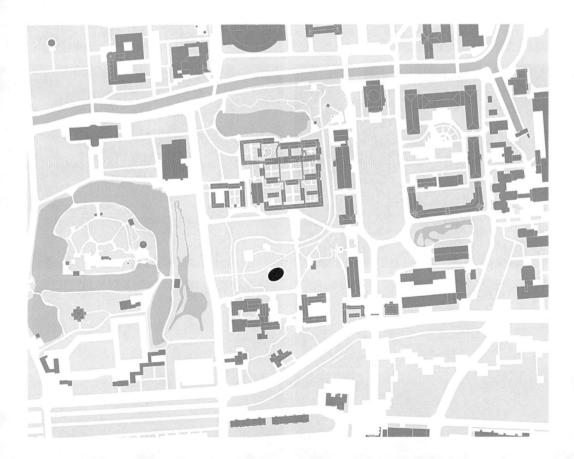

EARTH WOMB is a sinking building space for meditation shaped with the help of the surrounding and the dynamic interference of natural light. In the design, I take the human experience and the surrounding space as a part of the meditation, rather than simply define it as a device. The location of the scheme is located on the lawn of the campus, the womb like meditation space plunges into the ground, providing a sense of security to the meditator. At the same time, distant light permeate from the ground into the underground. With the change of time, the light and shadow of the underground space changes continuously and subtly. In addition, the gap

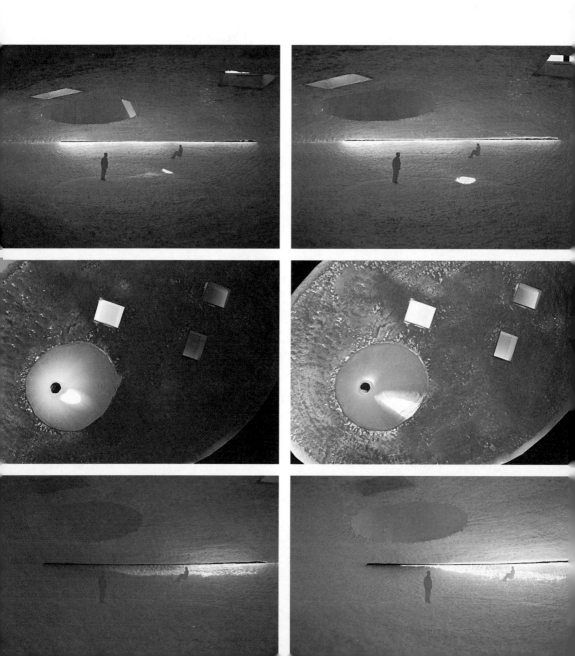

between the light collection tube on the top of the meditation space on the ground can project the activities on the ground to the underground through the changing light shadow. Therefore, the meditation activities are stimulated by natural light and the behavior of the people on the ground.

本页及对页图（一排）：直射光的变化。本页及对页图（二排）：仰视光影的变化。本页及对页图（三排）：缝隙渗入光的变化。

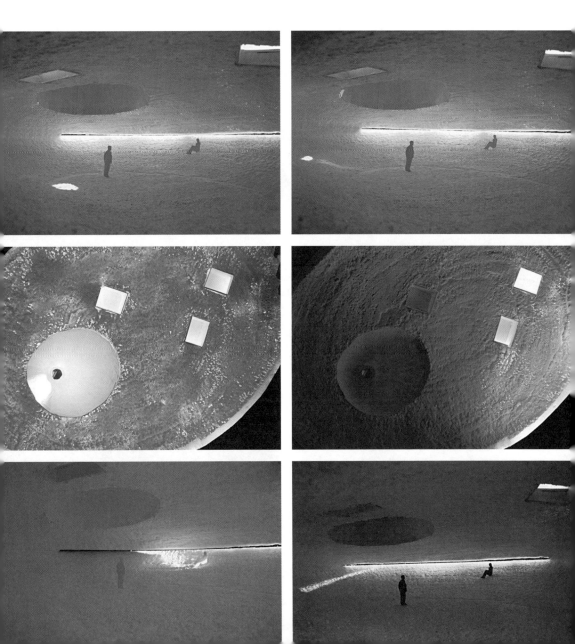

Open Design Studio 2014

教师点评

现代都市生活越来越拒绝"暗(darkness)"了。而我们的生命和身体,恰恰是从黑暗中来。暗具有某种特殊的力量,使我们的心境得以抽离于尘世,将感知和意识从单纯的视觉回归身体。秦正煜的这个设计,就是在探讨一个关于暗的空间的塑造。

一个微微隆起的穹顶,上方是一个小土丘,为这个繁忙城市中的路人提供一个歇脚的去处;而下方,却是一个寂静的"洞穴",一个暗的空间。通过仔细组织的几处孔隙,太阳光还是可以在不同的时间渗入空间,留下几抹亮的痕迹,又转瞬即逝,好像在应和着某种生命的节律。

Teacher's comments

Modern urbanism increasingly refuses "darkness". But our lives and bodies are just from the darkness. Darkness, endowed with certain special power, enables our state of mind to get rid of the ordinary world, and return the perception and consciousness from simple visual sense to the body. The design of Qin Zhengyu is discussing the creation of a dark space.

A slightly bulged dome, with a small mound in the upper side, provides a place for rest for passerby in the busy city; the lower side is a quiet "cave" and also a dark space. Through several carefully organized pores, the sunlight can also permeate into the space in different time, leaving some fleeting traces of light, which seems to reflect certain rhythm of life.

对页上图(从左至右):人与采光盒的互动。**对页下图(从上至下)**:人与主要采光筒的互动。地上与地下人的互动。**跨页**:主要室外场景。

光的呼吸/BREATH OF LIGHT

项目选址：清华大学小树林
功能定位：光的冥想空间
建筑面积：400m²
用地面积：400m²

方案设计：葛肇奇
指导教师：董功
完成时间：2014

方案旨在探索一个由自然光主导的冥想空间。地面上的采光筒朝向太阳轨迹上的不同方位。随着太阳的转动，地下的"宣纸灯"将被依次点亮。

外观上，低矮的底部承托起上方7只筒体，仿佛是场地上一个大型雕塑。行人首先会在墙中穿行，接着从一个狭长的坡道下至地下3m的空间。"茧"从屋顶垂下，光饱满充斥其中。人既可以坐在茧外地面小憩，也可顺着缓坡钻入茧内，可以躺视万神庙一般的天空片段。顶部采光筒开口方向不同，所以一天中不同的茧会在不同时刻达到最亮的状态，光影会呈现出明暗交替的时间性而非简单的线性移动，冥想的人仿佛成为了其中孕育着的生命。

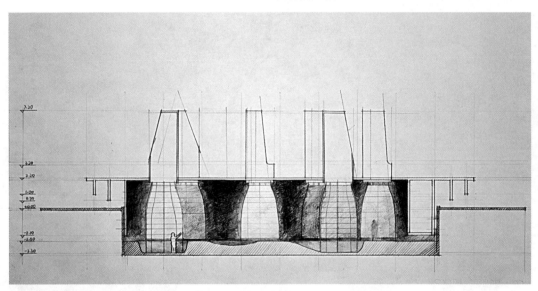

开篇：主要方案图。对页图（从上至下）：剖面图。立面图。本页上图：平面图。本页下图：总平面图。

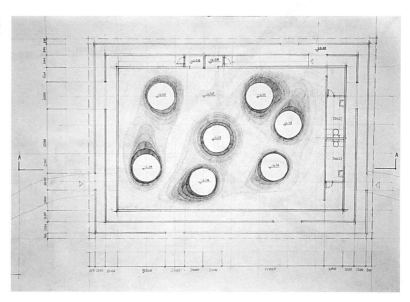

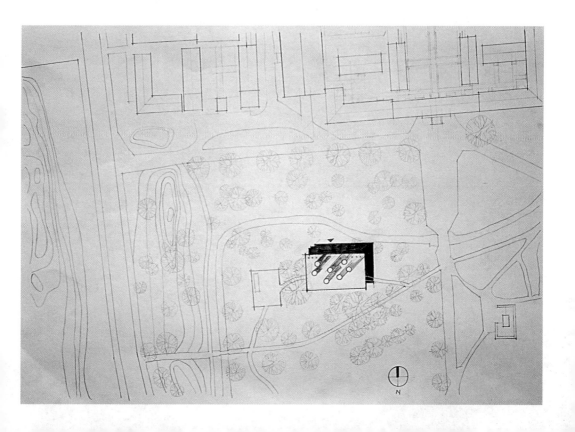

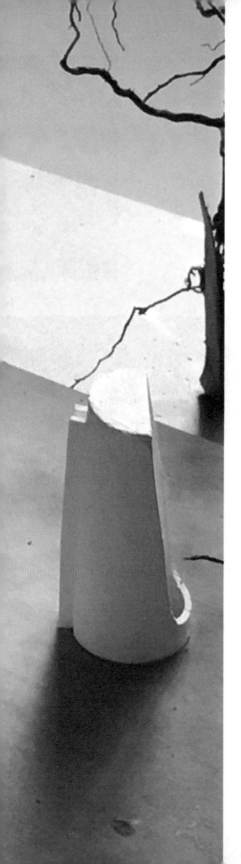

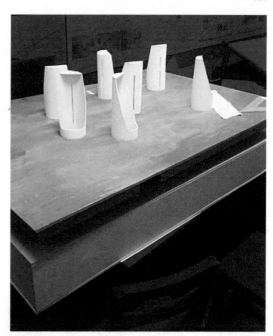

跨页图：小比例方案模型。**本页上图**：1:20光影实验模型。

The project aims to explore a meditation space that is dominated by natural light. The light collection tubes on the ground face different directions according to the solar path. As the sun rotates, the "Rice paper lamps" underground will be lit one by one. On the appearance, the lower bottom bears the lower 7 cylinders above, which is just like a large sculpture on the site. Pedestrians will first walk through the wall, and then go to the 3-meter underground space from a narrow ramp. The "cocoon" droops from the roof, the light is full of it. People can sit on the ground outside the cocoon to have a break, and can also plunge into the cocoon along the ramp, having a view of the sky fragment when lying down. The opening directions of the light collection cylinder on the top are different, so different cocoons in one day will reach the brightest state at different times, the shadow will present a alternating light and dark rhythm rather than simple linear movement, and the meditator seems to have a life nurtured therein.

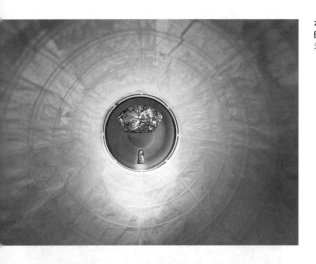

本页上图：特殊视角场景。**跨页上图**：建筑灰空间下。**跨页下图**：1:20 光影实验模型内部。

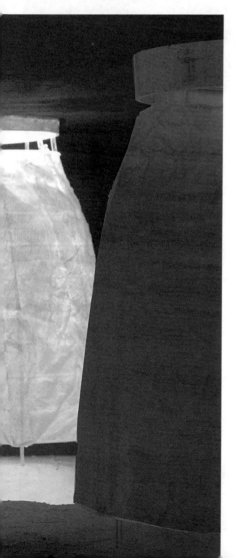

教师点评

设计者在制作案例分析模型的过程中,对"糊纸"材料的半透明性及其对光线的敏感产生兴趣,进而用其将人"包裹",形成空间单元,收集一天中来自不同角度的光。它们并置在一个下沉的、表面"柔软"、自由的幽暗空间,随时间流逝,各单元明暗交替。设计者通过模型和影像直观地呈现了自然光的魅力,抓到了他所理解的光所具有的特征。如果说密斯的通用空间是物理般明亮的、绝对自由的,那么,设计者创造的则是另一种意义上明亮的,甚至可以说是现象上充满光的通用空间。当整个昏暗空间被"跳跃"的光"茧"点亮,光在空间中真正成为主角。光的存在,以及人行为的趋光性,让绝对自由的空间变得相对自由。这个起点来自设计者匠人般手工的"糊纸",在现实中,真实的材料到底会是什么呢?

Teacher's comments

In the process of making case analysis model, the designer becomes interested in translucence of "papering" materials and its sensitivity to light, and uses it to "wrap" people, to form space unit and collect rays of light from different angles. They are jointly placed in a sunk and free dark space with "soft" surface. As time goes by, shining in the darkness alternates in each unit. Through model and image, the designer visually shows the charm of natural light, and grasps the characteristics of light he understands. If universal space of Mies is physically bright and absolutely free, universal space created by the designer is a kind of bright space in a different sense, or even a universal space full of light phenomenally. When the whole dark space is lightened by "jumping" deflossed cocoon, the light becomes the real protagonist in the space. The existence of light and phototaxis of human behaviors make the absolutely free space become relatively free. This starting point is from the craftsmanlike handmade "papering" of the designer. So in reality, what's the real material? (Written by Liang Chen)

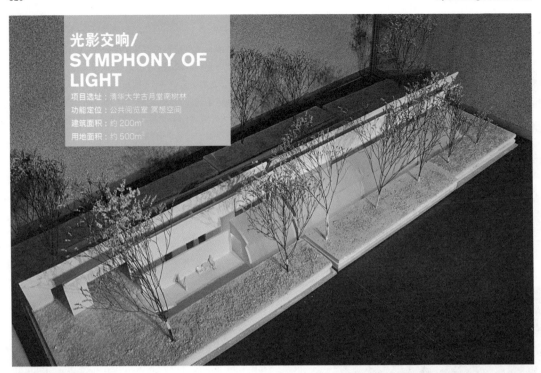

光影交响/
SYMPHONY OF LIGHT

项目选址：清华大学古月堂南树林
功能定位：公共阅览室 冥想空间
建筑面积：约 200m²
用地面积：约 500m²

方案设计：辛大卫
指导教师：董功
完成时间：2014

影是三维物体在二维平面的反映。通常来说，影所包含的信息要少于物体本身。但通过半透明材质、倾斜墙体和破碎不连续的表面，光影的维度被改变了，进而创造出一个使人被三维光影包围的空间。随着时间推移，光影变化呈现出多样性和差异性，内与外、虚与实的界限被打破，给予身处其中的人"冥想"的观感和氛围。

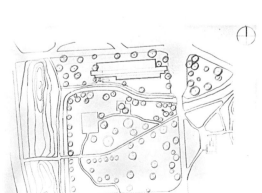

天·地·光·城市／SKY, EARTH, LIGHT, CITY

项目选址：五道口易初莲花超市斜对面空地
项目类型：冥想空间
建筑面积：280m²
用地面积：680m²

方案设计：张若诗
指导教师：董功
完成时间：2014

本设计意图在喧闹的城市中创造一个以"光"为媒介将人与天地"联通"的冥想空间。建筑位于五道口，由一个悬挑在两面墙中的方盒子和其中通向"天地"的光筒组成。身处其中，随着时间、天气、外部人流与车流活动的变化，自然光与人造光在整个空间中将人的感官体验无限放大，生命的微妙细节再一次彰显。城市成为了冥想空间的"画外音"，此时，虽耳畔喧闹，却内心宁静。在此，天、地、人、城市因光而生动、融合。

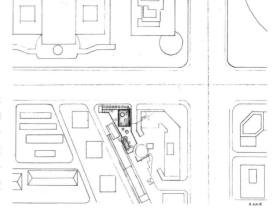

水·星·光/WATER, STARS, LIGHT

项目选址：清华大学古月堂南侧绿地
功能定位：公共休息亭
建筑面积：100m²

方案设计：赵萨日娜
指导教师：董功
完成时间：2014

方案立足于思考光与水的关系，以简洁的形态和细微的透光孔的设置塑造极度幽暗的环境，以黑暗来衬托并凸显光的效果，借此体现出一天当中的不同时刻下光照在水面上产生的不同效果，例如日出日落时光照贯通的绚烂和正午时的星空效果。

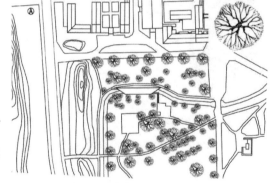

光的空间/SPACE OF LIGHT

项目选址：北京市海淀区中关村东路13号，华清嘉园（西区）小区内
项目类型：休息亭
建筑面积：208m²
用地面积：746m²

方案设计：杨宝诚
指导教师：董功
完成时间：2014

自然光作为人们日常生活中最广泛接触同时也是最易被忽略的元素，其纯净表达极具感染力；在地段中，社区的舒缓自然节奏与城市的急促人工节奏并置。方案通过建筑手法吸纳两方特性，借助光的现象予以再表达，在光的交融中创造感人空间。

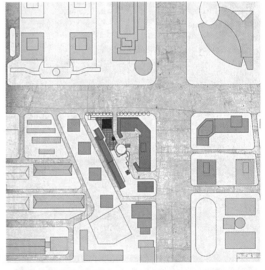

树之眼 / EYE OF THE TREE

项目选址：清华大学校友林
建筑面积：300m²

方案设计：周祎馨
指导教师：董功
完成时间：2014

方案试图以光为媒介，将建筑的时间感和空间感充分结合。通过剥夺建筑的功能性，设计方案以建筑空间对人的内心世界的影响为重点。不同形状高度的屋顶，缝隙，开口，将人的视野引向上空，又让光影从屋顶有节奏地洒落，激发起想象和幻想。强烈的明暗对比创造出空间的平静和存在。

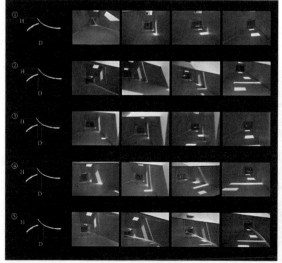

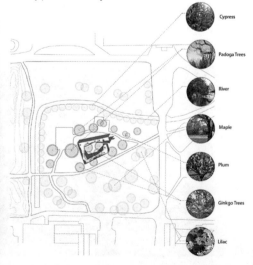

光的冥想空间 / LIGHT MEDITATION SPACE

项目选址：清华园甲所草坪
建筑面积：144m²

方案设计：左碧莹
指导教师：董功
完成时间：2014

设计的着眼点是场地上茂密的树林。设计概念是在树的四个不同高度层次上以光为媒介来体验树。

整个设计的空间体验过程也像一个生命的体验过程：从一个原始的初始状态，逐渐成长上升，其间遇到了困难黑暗，然后又突破了黑暗，取得了光明繁盛，最后穿过繁盛进入超然物外的境界。

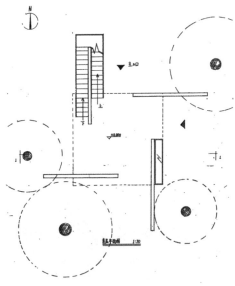

光的滤镜 / OPTICAL FILTER

项目选址：清华大学工字厅前
功能定位：冥想空间
建筑面积：300m²
用地面积：500m²

方案设计：唐宁
指导教师：董功
完成时间：2014

这个方案最初的灵感来源是日本的纸门：薄薄的一张纸就将斑斓的世界阻挡在外，留下最纯粹的光影。我所设计的屋顶便如同一扇纸门，深梁阻挡了人们的视线，却能让光影渗入、让树叶洒落，人们能够在此暂时脱离花花世界，回归平凡。

时·光·序列 / TIME, LIGHT AND SEQUENCE

项目选址：清华大学甲所
功能定位：冥想空间设计
建筑面积：158m²
用地面积：632m²

方案设计：李新新
指导教师：董功
完成时间：2014

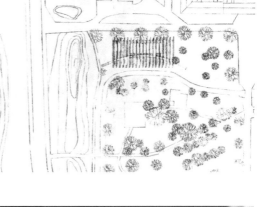

这次以光为研究对象的冥想空间设计，与其说是一次实践，不如说是一次对建筑本质的探问。自然光随时间变化，对空间产生不同的重要影响，设计者在设计过程中一直试图寻找光、空间与时间的关系，尽量从功能和形态的束缚中解脱出来，以一个思考者的角度创造一个理想的冥想空间。

致谢 Acknowledgements

2014年春季学期，清华建筑学院在校长、人事处和教务处的支持下，一次性聘任15 位业界优秀建筑师作为"校聘设计导师"，进行了开放式建筑设计教学的尝试，同年春季学期，有8 位作为指导老师参与了三年级建筑设计课教学，其余7 位则参加了设计评图。2015年春季学期，同样有8 位设计导师参与了设计课程教学，他们是马岩松、王昀、王辉、朱锫、齐欣、李虎、徐全胜、崔彤，其余7 位设计导师董功、胡越、梁井宇、华黎、邵韦平、李兴钢、张轲参加了设计评图，在此对15位设计导师的辛勤工作深表敬意。

在教学过程中，清华建筑学院大部分设计教师参与了开放式教学的中期评图及最终评图，他们与设计导师们进行了多方面的讨论及交流，并对开放式教学提出许多宝贵的意见，在此深表谢意。

本作品集的编辑过程中，所有选修这门课的学生参与了作品的排版工作，感谢同学们的努力；这里还要特别感谢杜頔康、李晓岸、罗丹、卢倩、丁立南、赵一舟等研究生们所做的编辑、翻译、校核等工作；最后感谢中国建筑工业出版社的编辑们为本书的出版进行了不懈的努力。

2016年7月